THE Mind IN THE Matrix

WHAT THE COMPLEXITY OF THE
UNIVERSE TELLS US ABOUT MEANING

To know how to wonder and question is the first step of the mind towards discovery.

Louis Pasteur

THE Mind IN THE Matrix

WHAT THE COMPLEXITY OF THE
UNIVERSE TELLS US ABOUT MEANING

ROBERT WILES
MB BS FRCGP FRACGP FACRRM

Information Press (Australasia)

To: Tim, Amy, Alex & Lucy

First Published in 2019
By Information Press (Australasia)
PO Box 159
Cooma. NSW. 2630
Australia
www.infopress.com.au

ISBN-13: 978-0-9875622-2-7

National Library of Australia
Cataloguing-in-Publication data is available from:
http://catalogue.nla.gov.au
Text design and typesetting: Derrick I Stone Design
Cover design: The Marketing Side
Cover photo: @tampatra TWENTY20
Printed by: Lightning Source

Foreword

Dr Robert Wiles, medico and polymath, has achieved something imaginative, unique and profound in this little atom-bomb of a book. He has translated complex knowledge and research in physics, quantum mechanics, mathematics and other fields to reveal that our universe and the mechanisms of all physical processes, including life, are digitally coded with embedded information.

Dr Wiles' great contribution to the field of human knowledge and spirituality is to allow that the unimaginably complex structure and functioning of our mathematically designed universe, (with its concomitant governing laws and ultra-precise settings) could not have materialised from nothing by accident. This is an impossibility. Therefore he has cut the Gordian knot regarding the greatest question of all, by revealing that after the 'Big Bang', the formation of the Universe, thence the Earth and appearance of life, were designed and defined by inconceivably complex information embedded in the laws and fabric of the universe.

Consequently, there had to have been a huge and unfathomable 'Mind' behind the Cosmos – which forces us to go beyond reductionism and materialism and acknowledge that our Information-based universe combines both 'material' and 'non-material' elements. The latter includes the incomprehensible Mind behind the designer, otherwise there is no meaning.

As Wiles therefore concludes in this original, exciting and ground-breaking work: because 'we live in a material universe that is underpinned by an intangible information fabric', the only rational and possible explanation for this *Universal Information Generator* (logically coming from an outside domain, and from outside time) is God. This then adds true meaning to our brief time here on Earth and, in Wiles' words, 'to a deeper reality.'

Professor Charles Massy
Ecologist and author of
Call of the Reed Warbler

Prologue

"Meaningless! Meaningless! ... Utterly meaningless! Everything is meaningless!"[1] These words that might have been uttered by any number of people today, were penned 3000 years ago by King Solomon as he contemplated life. Three millennia on, widespread anxiety about the future and the alarming suicide rates reflect a persisting lack of purpose and suggest that the thirst for meaning remains a critical issue. Meaning gives significance to everything we do so without it, our lives become pointless.

With our growing affluence and expanding leisure time there have been increasing numbers of people searching for meaning, which has been paralleled by a ten-fold increase in clinical depression since 1945.[2] The 2018 Mission Australia Youth Survey of 28,000 responders aged 15 to 19, identified mental health as the top issue for young people and it is alarming to note that this concern has doubled in the last three years.[3] Conversely, well-being is *positively* impacted when we perceive that our lives have meaning; however, meaning arises through information, so to understand meaning we must first investigate information.

It is now more than thirty years since the emergence of the Information Age but we still have only a very hazy understanding of the nature of information. R.Z. Sheppard outlined the problem in 1971when he wrote: "*In much the way that fish cannot conceptualize water or birds the air, man barely understands his infosphere.*"[4] The information sphere or infosphere is the total realm of information. Luciano Floridi, Professor of the Philosophy and Ethics of Information, University of Oxford observed: *"Work on the concept and hence the definition of information is still at the lamentable stage when disagreement affects even the way in which the problems themselves are provisionally phrased and framed."*[5] Therefore, the search for a definition and understanding needs to encompass all information, which maybe the IP packets whizzing around the internet but also the store of information within the DNA inside your cells.

As will be seen in Chapter 1, although no one can definitively say what information is, mankind has long recognised the *effects* of information. Physics Professor Paul Davies' expresses his struggle as he has grappled with this when he writes:

"Through my scientific work I have come to believe more and more strongly that the physical universe is put together with an ingenuity so astonishing that I cannot accept it merely as a brute fact. There must, it seems to me, to be a deeper level of explanation."[6]

That deeper level of explanation can be clearly understood through information that builds to create knowledge, then through wisdom provides meaning. As the Information Age has advanced, there has been a growing appreciation of the significance of information to the extent that we are changing the way we view the ultimate nature of reality, from a materialist standpoint, where physical objects and processes predominate, to an informational one.[7] A typical example is the chunky black telephone of my childhood that was tethered to the wall delivering information via voice, which has morphed into the sleek smart phone that is adorned with apps delivering applied information on the run. We perceived the analogue telephone primarily as a physical object but understand the smart phone as an information device.

We are also becoming more aware that we are constantly being bathed in vast quantities of this impalpable, intangible information. A typical example is the software that regulates every living organism through its DNA; although the information code is *non-material*, life cannot survive without it. However, to exert its influence within our physical universe, information requires material carriers which, in this case, we call chemicals. The

DNA molecule *carries* the code of life that provides meaningful instructions for the cell to function, while the molecules themselves are just meaningless substances, mindless actors that dance to the tune of information. In your mobile phone, the handset is the physical platform which carries the non-material apps that deliver the information. Note that only non-material information can provide meaning and purpose, in contrast to physical matter and energy, the other fundamental entities of the Universe, which do not. So despite previous assertions to the contrary, the observable evidence demonstrates that information is real and *radically* different.

The unexpected way that mathematics often describes our universe is examined in Chapter 2. Nearly all scientists agree that the Universe began at the "Big Bang" when our entire universe is considered to have materialised out of nothing. Many would suggest that there was no foresight or planning in this process, yet the Universe is described by exquisite mathematical statements and all known fundamental laws are found to be mathematical in form.[8] Some physicists now propose that this is because information specifies the laws of the Universe. Yet meaningful information cannot self-assemble, implying that there must have been an information generator or programmer to orchestrate the order described by the mathematical code that describes the Cosmos. Note that mathematics is just an abstract tool, a language, that also requires an information generator. The problem is that this is completely beyond the explanatory power of the Standard Model of Cosmology.

Chapter 3 investigates some of the more than 100 tightly defined parameters that are required for the Earth to support life as we know it; the Cosmological Constant must be specified to 137 decimal places! These previously unimagined constraints leave science at a loss to explain but it is clear that our universe is *finely tuned for life*. The only credible answer, as Nobel laureate Fred Hoyle remarked, is: *"that a super-intellect has monkeyed with physics"*.[9]

From the biosphere we drill down to living cells which are information rich and even the simplest cell is mind-bogglingly complex. Cells are tiny factories operated by microscopic machines driven by digital information. The Human Genome Project published in 2000 showed that most of our DNA is an instruction manual, which is unbelievably information dense, meaning that each cell is an information processor! Darwinian evolution can only operate through living organisms so cannot explain the appearance of the *first* cell. However, life required meaningful information to commence, so in addition to the previously identified *physical* information generator that created the Universe, there must have also been a *biological* information generator to write the code for that first microbe.

Chapter 5 explores the mind and life, which are universal to human beings and considered to be amongst our most precious possessions, yet they are two of the hardest concepts to understand in scientific terms. Similarly, consciousness and self-awareness persistently evade materialistic explanations. Eminent philosopher Charles Handy comments: *"...meaning is rooted in our consciousness, which cannot be coded or made into data. Nor can the virtues of beauty, truth, or love, which you recognize when you see them, but cannot adequately define. Love, trust, loyalty, and judgement—the essentials of our human relationships – are also immune to digital quantification. Trying to codify them is pointless."*[10] Despite our inability to digitise these qualities, computers do offer a very useful analogy to assist us. Once we appreciate that the hardware (our brain), the software (consciousness) and the information they process (our thoughts), are *not* the same entities, it is readily apparent that our brain (the physical) and our mind (an intangible) are different. Mind, including thought, memory, consciousness and emotion are more easily understood once we recognize that it is a non-material entity running on a physical platform. Together these elements describe a *Compound* Model of Mind which is the foundation for our appreciation of meaning.

The human brain is the most complex entity known to man and self-awareness is found within it alone. Thus human beings are unique within the Universe. However, scientific experiments over the past century have been unable to identify any particular place within the brain that self-awareness might reside. Instead, some scientists have proposed a holographic mechanism where mind is everywhere within the brain and yet nowhere in particular. Chapter 6 delves into the implications of these discoveries, showing that computer processing, operating within *Cyberspace* and our thoughts within *Mindspace* are not the same. Perhaps most surprisingly, we will discover that our non-material mind has the ability to determine the way our physical brain is networked, a feat completely beyond the capabilities of your computer.

In summary, we live in an exquisitely ordered but amazingly complex universe full of beauty and wonder which, as many scientists, including Einstein have pointed out, is completely unexpected. The reason is that the Cosmos is specified by an underlying substratum of pure information. After demonstrating that information generators are needed to explain the origin of the *physical* and *biological* components within our universe, the concluding chapter works through the significance of these findings, showing that an exclusively materialist description of the Universe cannot adequately explain the evidence before us. However, this extraordinary amount of information is observed to behave in one of four different ways. After the Tangible Domain (where we experience our physical existence), Cyberspace (computing) and Mindspace (thought), there is Quantum mechanics which occupies a fourth information domain. These are all explored and mapped, then unified to describe a Compound Model of the Universe. At times the details can seem overwhelming so to assist readers with differing levels of understanding of the sciences, information summarised in shaded tables and boxes is presented for those who want to dig deeper but can be skipped over if desired.

As we trace the origins of this immense quantity of information, the question that arises is why does it exist at all? This question has echoed down the centuries, as nothing is simpler and more probable than something.[11] To answer this question, we will be digging through the information within the chapters above. Professor Alistair McGrath of Oxford points out that science tells us much, but it can't detect meaning. Yet this doesn't mean there is no meaning to be discovered.[12] To overcome this inability, the scientific paradigm has been transitioning from an exclusively materialistic empirical way of thinking that struggles with meaning to include an information dimension that is rich in meaning. We will discover that any meaningful entity (such as life) requires a meaningful code (such as within DNA) and ALL meaningful code requires an intelligent source. This is the Golden Rule of Information. It is the tacit assumption behind SETI's sixty-year search for extra-terrestrial life. However, SETI appears to have been looking in the wrong place. The Space-Time fabric of our universe appears to depend upon an Information fabric that underlies it. The profound meaning this information affords our universe *necessitates* an intelligent origin: The Mind in the Matrix!

Robert Wiles
October 2019

Contents

In many respects, we are living during a time best described as the 'Mis-information Age'. *The Mind in The Matrix* changes all of that! Robert Wiles brilliantly presents an eye-opening analysis challenging antiquated notions relating to the origins of the Cosmos (and life itself) and replaces it with irrefutable data that actually makes scientific sense. The book is an astonishing read leading to the inevitable conclusion – Julie Andrews (Sound of Music) was right! *'Nothing comes from nothing, nothing ever could.'*

Gary Ezzo, Author
Parenting From the Tree of Life

This book by Robert Wiles will challenge your view of God's good creation as biology and physics intersect and collide in ways and at dimensions that few know, and even fewer understand. Dr Wiles explains with rare clarity for both the expert and novice alike how all of our universe is interconnected and interrelated in a vast complex matrix that not only calls out to a master creator but screams: '*I Am Who I Am*'.

Dr Kenneth Meyer
Engineering Electrical/ Electronics
Canberra Institute of Technology

Chapter 1

Searching for Meaning in the Information Age

Modern times have delivered most people in the western world unparalleled prosperity compared with even their grandparents. This prosperity has been accompanied by rapidly expanding knowledge yet our sense of meaning and fulfilment seem to be shrinking just as fast, if not faster which is leaving a void we long to fill. Dr Alister McGrath comments: *"We are indeed meaning-seeking creatures. Modern psychology points to there being a universal human sense of a need for meaning...We want to understand our experience and to feel that our lives have significance and purpose."*[1] McGrath points to a massive literature, not the least of which is: *Man's Search for Meaning* written in 1946 by the Austrian psychiatrist, Viktor Frankl. Professor John Lennox of Oxford observes: *"[Frankl] believed that the primary human motivational force is the search for meaning. The fact that millions of copies of his book have been sold in many languages demonstrates the aching void that exists in the human heart."*[2] This longing for answers has not diminished with the advent of the Information Age. Despite having access to more data more quickly and more easily, the suicide rates continue to rise and anxiety is becoming epidemic. To unpack this, we will use the tool of our age, Information, to make sense of what we observe.

However, if meaning is daunting, information has its challenges too. Even though the Information Age began over 30 years ago, information is still a mystery.[3] Anthropologist, Terrence W. Deacon writes: *"...although we use the concept of information almost daily without confusion, and we build machinery (computers) and network systems to move, analyse, and store it, I believe that we still do not really know what* **it** *is."*[4]

Information is different to most other things we encounter. In the past, information has often been viewed as a concept or a useful tool to analyse a situation but information is more than this. It is commonly said that *"information is power"* but why is it powerful? The reason is that it allows us to make sense of the world around us, to make predictions that help us to anticipate the future and make the most use of future opportunities, such as knowing when and how to apply for a job, make your next investment or even just catch a fish for dinner! So this chapter explores the characteristics of information that impart meaning.

The Information Age

Previous civilisations *used* information for survival, such as knowing the best way to grow crops for food; hence the Agricultural Age that lasted until the invention of the steam engine in the mid-1700's. This paved the way to the Industrial Revolution when the effective use of energy supplanted manual manipulation of materials as the highest-value pursuit and determined the winners under this new set of rules. However, the World continued to change rapidly. In 1837, Morse code appeared, followed by the telephone in 1876, radio in 1895 and television in 1926 laying the foundations for a communications paradigm which we entered after World War II. The World began to shrink as the transmission of information commanded a higher value than the utilisation of energy.

Hard on the heels of the Communications Era came the Information Age. With the launch of the IBM personal computer in 1981, computers became user-friendly to business and the individual. The highest value commodity now became the collection, storage and retrieval of data. When that data was ordered it became information. It was then tested to create knowledge and contextualised to create wisdom. Re-contextualisation led to innovation and the dominance of the communications companies was eclipsed by the computing industry. However, the history does not end there. The automated analysis of rapidly accumulating data in the Cloud by ever more efficient search engines, such as Google, marks the emergence of the Knowledge Revolution. The Age of Robotics is yet to come, where new and novel information will be analysed and applied without direct human input.

Just a Concept

We encounter information every day; but what is it? Many people comment that information is *"just a concept"*. They consider that it is a useful way of thinking about the way we see our world but that is all. However, I discovered that information was a separate entity after I purchased my first computer in 1983. When I first switched on my beautiful new device, I was annoyed to find that it was useless until I purchased extra programs and installed them. I had to buy software in the form of an operating system (MS DOS at that time) and applications programs, such as a word-processing program as well. Once these were installed it became a very useful machine. But after the software was installed, the computer was no wider, deeper, longer or heavier than the hardware alone! There is no doubt that something of value was transferred from the floppy discs and installed onto the computer (try using a computer without the software) and most software packages involve many hours of intellectual pursuit to write. This is why our society recognises the value of information/knowledge by allowing it to be copyright protected and the special term 'intellectual property' has been coined to describe it. However, without the software packages the computer is useless as an information processor. It can be seen therefore, that a distinct entity, (contained within the software), that is more than the disk that carried the information, was installed onto the computer. If the information was *"just a concept"* then the information alone could not have affected the outcome, which was to allow the computer to process information. Also note that information is not energy either. Energy can be used to transmit information such as the beam of light from a lighthouse but energy is merely the carrier of the information. Thus *information has no substance (mass) but it is real* as it changes function and outcomes. Information is a discrete (yet non-material) entity in its own right and must be considered separately from matter and energy (which are actually different forms of the same thing). So, there are only two fundamental entities in the Universe: matter/energy which is material and *information* that is not.

Hardware, Software and Information

Our principal tool for manipulating and managing information is the computer. The box or 'hardware' is indispensable as the physical platform on which the software runs. Software is the set of instructions (information commands) which tell the hardware what to do. Hardware is the physical apparatus that allows the intangible information, carried by the software, to operate in the physical (material) domain. As the hardware and software have to work together, they must complement each other. An example would be the difference between a Mac and a PC. Trying to run the software specifically written for one family of computer hardware on the other type will not work efficiently (if at all). Although the DVD (the vehicle) carrying the software instructions may be identical for both types of software and the resultant program that runs on the relevant computer appears almost identical on the screen, the instructions must be specific to each type of hardware. And yet, although the software written for each type of computer is unique to that platform, the underlying message that is being conveyed by the respective software packages is identical! Therefore, we must distinguish not only between the hardware and the software but also between the software and the message or information that is being carried by the software. This is the case in biological systems too. Each species has its own unique genetic code which specifies growth, shape and activity, yet behaviour can be almost identical. In addition, as the software is unique to each piece of hardware, it must have developed in parallel with the physical body of the living organism in which it runs. But which came first, the software, the hardware or the information? Note that each one is useless without the other two!

Thus, information is more than just the medium (hardware and software) that carries the message. This was classically seen in movies, where the bearer of bad news tried to distance themselves from the information they were carrying by saying: *"Don't shoot me, I'm only the messenger!"* In this case, the person is the 'hardware', the language they speak is the 'software' and the information is the message they bring. Some common examples are illustrated in Table 1.1.

	Computer	Novel	School	Music
Hardware	Box & screen	Paper	Whiteboard & marker	Paper Score
Software	Programmes	Language	Syllabus	Staves
Information	Data	Story	Knowledge	Music

Table 1.1: Examples of Hardware, Software and Information

Quality versus Quantity

In 1948, Claude Shannon described how packets of information, such as electronic signals in a telephone network, might be counted which became known as the Shannon Theory of Information. This is also called *syntactic information.*[5] However, Luciano Floridi comments: *"The term 'information theory' is an appealing but unfortunate label, which continues to cause endless misunderstandings. Shannon came to regret its widespread popularity..."*[6] Floridi prefers the term: 'The Mathematical Theory of Communication' or MTC which *"...is the theory that lies behind any phenomenon involving data encoding and transmission."*[6] MTC or Shannon theory is good for measuring the volume or quantity of information but is incapable of measuring the meaning or worth of information. Information which contains meaning is called *semantic information*. It is possible to have a large amount of data which is meaningless and hence, useless, so any complete measure of information must include both the quality as well as the quantity of the information. This is supremely important in biological systems. To illustrate, imagine a string of DNA with an endlessly repeating segment and compare that with the genome of an amoeba that is living. Thus, more is *not* necessarily better. The problem is that currently there is no way to mathematically measure the meaning or quality of a piece of information.

The Infosphere

Information increased rapidly over the 20th Century, to the extent that we are now drowning in it. Yet scientists have known for hundreds of years that information existed throughout the Universe and that it is arranged layer upon layer and extensively networked. Floridi writes: *"...we are not standalone entities, but rather interconnected informational organisms..., sharing with biological agents and engineered artefacts a global environment ultimately made of information, the infosphere. This is the informational environment constituted by all informational processes, services and entities..."*[7] A living cell is a good example. In 1998, Bruce Alberts, President of the US National Academy of Sciences described how cells can be viewed as factories that contain elaborate networks of interlocking assembly lines composed of large protein machines that contain highly coordinated moving parts to conduct the business of the cell.[8] Each nano-assembly involves another layer of interlocking information.

Information Transfer

The transmission of matter and energy is usually obvious but information encoded within them may go unseen. John Lennox writes: *"...information is invisible. The carriers of information may well be visible - like paper and writing, smoke signals, television screens or DNA-but the information itself is invisible"*.[9] Thus, to detect the information, the method of storage or the carrier of the information needs to be recognised, so that the information *itself* can be identified (like radio waves).

Information is also transferable over time; if it was not, we would not know of the exploits of Julius Caesar. This can only happen, however, if there is effective transmission of the information via its carriers. During the Dark Ages in Europe, a lot of the ancient Greek and Roman writings were lost to European scholars until they were rediscovered during the Renaissance in the Middle East and Spain, where they had been preserved by Muslim scholars. Without these remote libraries we would not know about much of the ancient world.

Complexity

Data becomes information when a relationship is specified. Complexity is structured information and is integral to meaning.[10] Structured information requires further relationships to be specified, to link those pieces of information in that particular way, which in turn requires activity to occur. As linking information requires effort, it is an active, not a passive process and from the Second Law of Thermodynamics it can be deduced that in an isolated system, information carriers (and hence functional information) cannot build itself spontaneously.

To measure the complexity of a structure, an arbitrary value may be assigned to each link or relationship which I have named an 'infobit'. Living organisms are made up mostly of proteins which are folded strings of component parts called amino acids. A typical protein made from 250 amino acid units has a *minimum* of 250 infobits specifying its sequence of amino acids and *at least* another 250 infobits specifying its folded three-dimensional structure which equals a minimum of 500 infobits. If the information in the human genome with its 3 billion (10^9) base-pairs or the thousand trillion, trillion molecules in the human body,[11] there are 10^{20} molecules. Expressed in infobits there are, therefore, potentially between 10^9 and 10^{20} infobits in each human being, an exceedingly vast amount of information. There is an information potential stored in each meaningfully related piece of information, which would appear to be lost if that relationship between the linked bits is disrupted. Further, this complexity comes at a cost; both an initial cost to build it and an ongoing cost to maintain it.

The Generation of Information

So information is a real entity that can enhance outcomes when it is introduced to a system (such as software in a computer) or prevent outcomes when it is removed (such as the inability of a cell to duplicate without its DNA) but where does the information come from? The model of the Universe that is accepted by most scientists today states that everything began at the 'Big Bang', before which there was nothing. If this is correct, at some stage after the formation of the Universe, the information that we now observe in the Cosmos must have been generated. How did this occur - for the Universe generally and for living cells specifically? In information terms there are only three possibilities. Either:

A. the coding of the information occurred by chance, which required no additional input; or,
B. a law of the Universe required that the coding of the information *had* to happen, which represents intrinsic information within the law that caused it to happen or,
C. external coding of the information occurred; new information was introduced into the system. These alternatives are illustrated in Figure 1.1.

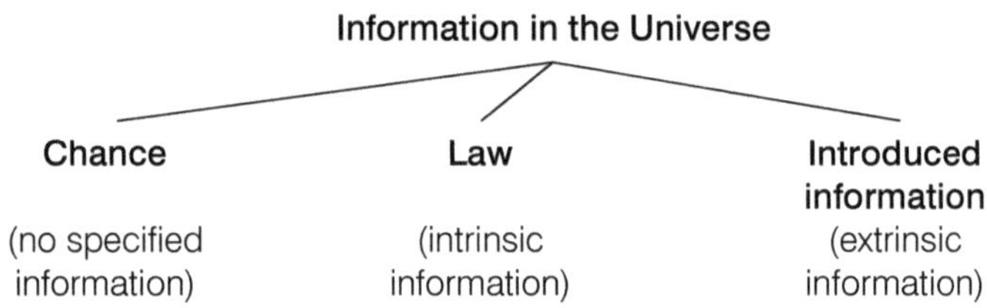

Figure 1.1: The Generation of Information

There are only three choices. This illustration uses the example of the formation of the Universe but it could equally be any other information containing entity such as the first cell.

So, can information be generated by chance as T.H. Huxley asserted in his Oxford debate with Bishop Wilberforce in 1860; equivalent to monkeys bashing on typewriters?[12] The answer is a qualified yes: *meaningless* (Shannon or syntactic) information can be generated by random means but *meaningful* (semantic) information cannot. There is some useless information in the random sequence of the letters typed by the monkeys but they lack any functionality. The meagre amount of syntactic information generated by chance cannot explain the immensely intricate, meaningful information that is observed in the laws of the Universe or a living cell. Chance produces disorder. Order that produces functionality is not created by the accidental selection of a portion of chaotic randomness.

However, law is no help either. As Philip Johnson, Emeritus Professor of Law, Berkeley observed: *"Law produces…simple repetitive patterns, as in crystals, but they can't produce the complex, specified sequences by which the nucleotides of DNA code for proteins…"*[13] Indeed, information in living things is novel, which is the requirement for a working cell. That leaves introduced information as the only possible explanation.

The Identification and Interpretation of Information

Information is ubiquitous but often silent. Our five senses are physical detectors so identifying and interpreting information can be challenging. X-rays were unknown to Isaac Newton, yet they exist and we know they are real. We just had to wait for the appropriate technology to be discovered for them to become detectable. Further, it is not just the identification of information that is challenging but so is its interpretation as physical tools are unhelpful. Former Cambridge Professor of Physics, Paul Davies commented: *"No physical laws will define information."*[14] The reason is that information is non-material and therefore requires different tools to analyse it (such as an information survey or audit –equivalent to counting the infobits within the entity).

The Significance of Philosophy

Although information was initially used for practical physical purposes, such as making more efficient ploughs or crafting better swords, in time, the ancient Greeks recognized the usefulness of studying the world around us and they began to seek knowledge for its own sake. When that knowledge, which is applied information, became regularised and systematised it became known as 'philosophy' (literally, 'the love of knowledge'). Philosophy is the study of the causes and behaviour of all things, both material and abstract. Philosophy establishes the fundamental rules for the accountability of all human activities and has as its aim: the pursuit of understanding (from which we derive meaning). This includes the study of information or ideas that cannot readily be tested or verified by repeated experimentation, such as origins. In contrast, empirical science is the study of information that has been submitted to repeated experimentation or direct observation with consistent results.

The Importance of Science

Double Nobel laureate, Linus Pauling stated: *"Science... is the search for the truth."*[15] Which begs the question - what is truth? From the above, it can be seen that philosophy is the father of science from which it developed and with which it should be consistent. 'Scientific method' derives from the observation and logic required by 'philosophic method'. These principles keep science objective and accountable. Science works well with tangible objects such as solids, liquids and gases, because experiments can be conducted, results can be demonstrated and the process can be repeated to confirm outcomes. Conclusions can be rigorous with tangible entities but empirical science struggles with intangibles such as thoughts and feelings. Yet we know that thoughts and feelings exist because we experience them every day and they change outcomes. However, it is one thing to experience something and it is quite another to understand and give an explanation. One current concept of science maintains that there is a *physical* cause for every effect; this is known as 'Materialism'. So where does that leave thoughts and feelings? Science is seen by many as the final arbiter in distinguishing truth from error, yet it struggles with intangible, non-material entities such as information.

Science is a method or a tool rather than an objective in its own right. Better expressed as the *'discipline of Science'*, it acknowledges those facts or ideas that can be explained by *'scientific method'*. Thomas Fowler and Daniel Kuebler wrote: *"...it is useful to distinguish two types or classes of explanations, rational and scientific. A* rational explanation *is one that uses some type of plausible reasoning from premises to explain an observed phenomenon or state of affairs."*[16] Much of philosophy is in this category, as it does not lend itself to experimental testing. Fowler and Kuebler note however that: *"This does not mean that* [rational arguments] *are therefore inferior or incorrect,...nor does it mean that they are exempt from scrutiny...*Scientific explanations *are a specific subset of rational explanations that are based upon the scientific method...Both the ability to make testable predictions and* *the possibility of being refuted are considered essential to scientific explanation."*[16]

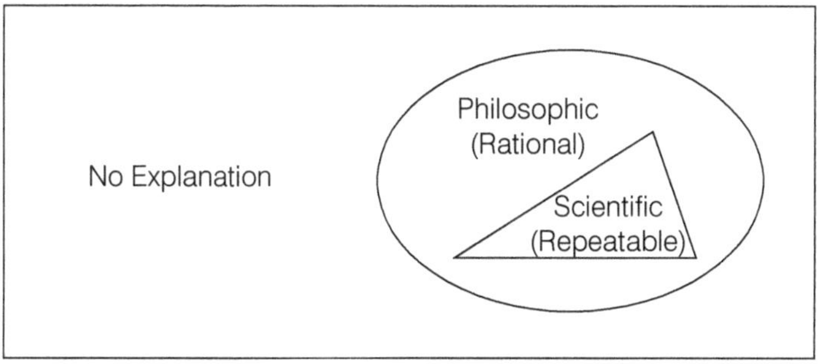

Figure 1.2: Types of Explanation

Philosophy involves the study of rational (logical) arguments which are consistent with observation. Science is a subset of philosophy, where the explanations are not only consistent with the evidence but testable, repeatable and falsifiable.

If a theory cannot be tested, then it ceases to come under the umbrella of science and is embraced by philosophy. Fowler and Kuebler reiterate: *"One central criterion of a scientific hypothesis is that the predictions it makes must be* falsifiable; *unfalsifiable predictions cannot be meaningfully verified by experimental tests, as no test can disprove them. In the case of evolution, for instance, Darwin told investigators to gather more fossils and they would find that most of the fossil record would consist of transitional forms....The key point is that investigators* are told to look somewhere they have not looked before, for something they have not seen before."[17]

It is interesting to note that this key prediction of Darwinian Theory continues to be vigorously debated, with Neo-Darwinians contending that it has been well satisfied by new fossil finds and their opponents contending that it has not. As science relies upon the premise that some events happen reproducibly and predictably and some do not, the implication is that some conclusions are 'right' and some are 'wrong'. On Earth, water always runs downhill due to gravity. This is an easily recognizable observation but unfortunately, not all situations are as clearly defined, as will be seen when Pasteur meets Darwin in Chapter 4. Either Pasteur *or* Darwin was right but both cannot be correct simultaneously.

Do not imagine that this is a simple process. There is a risk that evidence can be contorted to support a theory, rather than the theory being designed to explain the evidence collected through observation and experimentation. To complicate matters even further, human beings are not robots, lacking feelings or without preconceived beliefs and so, even the most devout scientist will have a bias due to their worldview, no matter how carefully they try to exclude it. The other problem is that the stronger our feelings or preconceived beliefs, the harder it is to exclude personal bias and there is no topic in science that is more hotly contested than origins and evolution. So discerning truth from error under these conditions is fraught with risk. When the debate leaves academic circles and enters the public domain where the rules are less demanding, the waters become even murkier as Post-modernism regards truth as a personal choice rather than requiring adherence to rigorous protocols or meeting independent standards.

Thus, science is a tool, a system of analysis, where the accuracy of evidence and the rigour of the logic are equally as important as the result. This process should lead us to precise and enduring results that are shown to be correct over the passage of time. The best measure of true data is that it is still predicting accurate outcomes and is, hence, considered 'correct', even after the passage of hundreds or even thousands of years.

So what is Information?

Floridi explains: *"Donald MacCrimmon MacKay...wrote that 'information is a distinction that makes a difference'...* [hence] *complete erasure of all data can be achieved only by the elimination of all possible differences."*[18] However most importantly, Floridi points out that not all information is the same. Although Shannon's (MTC) Theory suggests that the classic monkey randomly pushing typewriter keys is producing a lot of 'information' (actually static), this theory is about the volume of data rather than its semantic meaning. Although the information is not completely meaningless, it is not yet meaningful and since [information - meaning = data], Floridi suggests that the 'Mathematical Theory of data Communication' is a more appropriate description of this branch of probability theory than 'information theory'. Floridi cites Warren Weaver, co-author with Shannon of the MTC, who remarked: *"...The mathematical theory of communication deals with the carriers of information, symbols and signals, not with information itself."*[19] Thus, MTC describes or measures the mechanism of data communication within a physical carrier or channel whereas semantics measure meaning, (the message), independently of its channel.

Now that we have some understanding of the characteristics of information, we can begin to classify it. Information is an entity that may or may not possess meaning, which can take many forms. It can be non-functional, such as interference during a broadcast or functional such as the beam from a lighthouse. The many faces of information will be assessed as we meet them through the book. A brief overview is presented in Figure 1.3:

Information

Lacking meaning	**Meaningful**
[Conveys data]	[Conveys a message]
Functional	Purposeful
Quantitative	Qualitative
Stochastic (statistical) = MTC	Semantic (thematic)
Mathematically manageable (Shannon)	Indescribable mathematically
Can form naturally	Must be generated intelligently

Figure 1.3: Types of Information

Information that lacks meaning conveys data that may be described mathematically by the MTC or Shannon Theory of information and may form through the random operation of laws within our physical universe, such as sand dunes. Meaningful information is purposeful and conveys a message that must be generated by an intelligent entity, consider the Rosetta stone.[20]

Meaning derives from the Non-material

As the Universe is constructed from information, it is not surprising that wherever one looks, there is information. In addition, the deeper one looks the more information there is to be found due to networking and multi-layering. However, the information is often inconspicuous as it is intangible and so it often goes unnoticed. The obscurity of information despite its profound significance is a universal paradox. However, by refocusing on the *information content*, in contrast to the vehicles that transmit the information, we can gain a greater understanding of ourselves and our Universe. This distinction was illustrated by Nobel laureate Roger Sperry who stated: *"The meaning of the message will not be found in the physics and chemistry of the paper and ink"*.[21] It is the meaning conveyed by the underlying information that is valuable; the information cannot be explained in material terms.[22]

Chapter 2

Our Mathematical Universe

The Big Bang

The more we discover about the Universe, the more extraordinary it becomes. The Universe began with a burst of intense light - the 'Big Bang' hypothesis.[1] This model suggests that all the matter in the universe began as the size of a pin head where it was compressed so tightly that all matter was converted into energy. This theory envisions that all matter and energy within the Universe appeared from *nothing* in the first second. The Universe was then thought to have momentarily expanded very rapidly during the 'inflationary period' after which it slowed down. So how was all that energy and matter generated out of nothing and once it formed, what caused it to adopt the pattern that we call the Cosmos?

The only way to explain this is to infer that some process 'nudged' the infant universe into its current configuration, along with its laws and precise settings. In other words, there must have been information which came with the matter and energy that defined the universal laws and hence explains the resultant form of the Universe. Einstein's Law of Special Relativity states that matter and energy can neither be created nor destroyed but only converted from one to another ($E=mc^2$). So it would be more logical to propose that the matter and energy came from another place rather than out of nothing and rather than a chaotic, random explosion, the introduction of information with that first 'burst of light' would more logically explain the universe we know today.

Another unexpected finding is that the rate of formation of the Universe must be tightly controlled for life to be possible. If the universe expanded too rapidly, matter would disperse so that it could not form galaxies. On the other hand, if it expanded too slowly, matter would not spread out sufficiently to prevent it collapsing into one giant mass again.[2]

However, this notion of a finite universe that had a beginning is relatively recent. It was not until 1965 when Arno Penzias and Robert Wilson discovered low level energy readings that are thought to be the remnant of the energy released at the formation of the Universe (called the 'Cosmic Microwave Background Radiation') that cosmologists generally became satisfied that there was indeed sufficient evidence to demonstrate a finite universe.[3] This Big Bang model for the formation of the Universe has become known as the *Standard Model of Cosmology*.

Now despite being called the 'Big Bang Theory', its name can be deceptive. Physics Professor Jim Al-Khalili writes: *"...the big bang was not like a supernova explosion with all the matter flying away from a central point. The expansion of the universe is a stretching of space itself, with the matter embedded within space and carried along for the ride."*[4] So, the Big Bang is not so much a 'big bang' as a 'sudden super searing seismic stretch'. Al-Khalili likens it to a sheet of rubber that is stretched further and further and as it is stretched, the galaxies spread out with it;[5] but there are laws that govern this process and hence there must be imbedded information, which needs an explanation.

Our place in the Universe

We live in a truly spectacular universe! There are many things that make it extraordinary and the first and most obvious are its enormous size and dynamics. Our home planet, Earth, spins around our star, the Sun, at 107,000 kilometres per hour and our solar system spins around the centre of our galaxy at 914,000 kilometres per hour.[6] Our home galaxy, the Milky Way is estimated to have 100 billion stars and to cross it at the speed of light (the fastest speed attainable within our physical Universe according to

Einstein's equations) would take 100,000 years. The Milky Way is grouped with nineteen other neighbouring galaxies in the 'local cluster' but over 1 billion galaxies like our own can be observed, each with 100 billion stars, so the stars in the known universe exceed 10^{23} and to cross the known universe would take 13 billion light years![7]

Figure 2.1: The Milky Way *[Reproduced from Wikimedia Commons© Jon Lomberg.]*
The Sun with our solar system can be seen lying between two of the spiral arms where it is less likely to be disturbed by a supernova explosion.

Information and the Universe

So what is known about the information embedded in our Universe? The answer is: Quite a lot. Four fundamental forces cause the Universe to perform its cosmic dance: the strong nuclear force, the weak force, electromagnetism, and gravitation. These range over 40 orders of magnitude; some repulse while others attract. The mix of forces is both remarkable and complex.[8] All of these forces act in predictable and precisely defined ways. We use them to make predictions, such as: the timing of eclipses and the behaviour of laser beams which can be determined mathematically. They are extremely reliable, even over thousands of years (and as far as we can tell), across the entire Universe. But what compels these forces to unswervingly behave the way they do and most surprisingly, why is mathematics able to describe the behaviour of so much of the Universe? The premise is that there is embedded information that determines their behaviour. The behaviour of water molecules is very predictable because the fundamental forces act the same way each time an experiment involving steam, water or ice is performed. This is because the information defining the nature of these forces is fixed and does not change. It is because these forces are consistent that we can reliably predict not only that the Sun will rise tomorrow but also the time and the position in which the Sun will rise. The fundamental forces are summarised in Table 2.1.

Force	Effect	Energy	Carrier
Gravitation	Determines the orbits of the planets (cosmology)	Hydro-electric & tidal power	(Graviton)
Strong (nuclear) force	Holds protons & neutrons in nuclei, preventing them exploding (physics)	Nuclear fusion & the hydrogen bomb	Gluon
Weak (nuclear) force	Determines radioactive decay (physics)	Nuclear fission & the atom bomb	W & Z bosons
Electromagnetic force	Determines the electron shells (chemistry)	Chemical explosions (such as dynamite)	Photon

Table 2.1: The Fundamental Forces[9,10]

These four forces define the Universe however they are only the tip of the iceberg. In 2016, to understand the mathematical relationships better, the *L-Functions and Modular Forms Database* was launched containing data from over 80 mathematicians in 12 countries. It is an online catalogue of millions of mathematical objects and how they are related that on paper would fill 5 million pages.[11] The extent and intricacy of the mathematics that underpins the Universe is truly astonishing!

Mathematics, Physics and Information

John Lennox explores the relationship between mathematics and physics by quoting Eugene Wigner, Nobel laureate in physics, who wrote: "*'The enormous usefulness of mathematics in the natural sciences is something bordering on the mysterious, and there is no rational explanation for it...it is an article of faith.' The relationship between mathematics and physics goes very deep and it is very hard to think of it as some random accident. Professor of mathematics Sir Roger Penrose FRS, whose understanding of that relationship is unquestioned, has this to say about it: 'It is hard for me to believe...that such SUPERB theories could have arisen merely by some random natural selection of ideas leaving only the good ones as survivors. The good ones are simply much too good to be the survivors of ideas that have arisen in that random way. There must, indeed, be some deep underlying reason for the accord between mathematics and physics.'*"[12] It is now becoming clear that the reason that there is an accord is because information defines the relationships. Paul Davies describes the traditional relationship between mathematics, physics, and information as:

Mathematics → Physics → Information

According to this orthodox view, mathematical relationships are the basis of our existence. The physical world is an expression of these mathematical relationships, whereas information is a secondary, or derived, concept that characterises specific states of matter (such as an electron spin being up or down). However, Davies lays out an alternative view where information is regarded as the primary entity from which physical reality is built. This represents a radical shift in worldview, thus:

Information → Laws of physics → Matter

Davies points out that, after all, the laws of physics are informational statements: they tell us something about the way the physical world operates.[13] Thus information (albeit expressed in the language of mathematics) preceded the physics. Professor Edgar Andrews comments that the physics is only a description of *what* we observe in the Universe, it cannot explain *why* the Universe is the way it is.[14] It is the underlying information that gives the Universe its properties and explains why the physics works.

Some of these properties are summarised in Table 2.2.

	Properties	Effects if greater	Effects if smaller
1	Charge on an electron	Atoms are unstable[15]	Atoms are unstable[15]
2	Gravitational force coupling constant	Universe would have collapsed after the 'Big Bang'[15]	Matter disperses too rapidly for stars & galaxies to form[15]
3	Electromagnetic force coupling constant (Tolerance 4%)[16]	Electrons bind too tightly to bond atoms: water cannot form[15]	Electrons bind too loosely & water would not form[15]
4	Strong nuclear force coupling constant (tolerance 0.5%)[16]	Few light elements (like hydrogen) form & none heavier than iron[15]	Protons/neutrons do not bind; Cosmos is mostly hydrogen; fewer stable elements[15]
5	Weak nuclear force coupling constant	There would be no helium or carbon[16]	Universe would be all helium & no stars[16]
6	Cosmological constant (Anti-gravity force)	Universe expands too rapidly for galaxies & stars to form[16]	Universe would not expand after the 'Big Bang'[16]
7	Properties of water	If ice was denser than water the seas would freeze[15]	Water cycle depends on water vaporising at lower temps[15]
8	Carbon/oxygen nuclear Resonance (Hoyle)	Most carbon would be converted to oxygen &/or neon[17]	Insufficient carbon would form from beryllium[17]
9	Proton to electron mass ratio	Destabilises atoms: protons decay to neutrons if mass 0.2% larger[16]	Molecules do not form and life is impossible[15]
10	Proton to neutron mass ratio (Tolerance 0.1%)[18]	Atoms are unstable	Fusion in stars ceases depriving life of energy[18]

Table 2.2: Fundamental Properties of the Universe
To exist, our universe depends on many precisely defined characteristics (information).

The Laws of the Universe

The list of tightly specified parameters in Table 2.2 is backed up by an equally impressive collection of elegant mathematical laws which describe the functioning of our Universe. The overwhelming question is: Why do the laws work and how did they 'organise' in the first place? The fact that we recognise a discipline called 'science' testifies to the fact that the universe is subject to stable physical laws that have operated since its inception.[19]

So, what are these laws? Edgar Andrews explains: *"The laws of nature constitute unchanging reality whereas the laws of science are our frequently imperfect attempts to describe reality. It is the business of science to discover laws that describe, more or less accurately, how the universe works.... But...science...would [not] be possible without the unchanging substratum of natural law - the ultimate physical reality that science strives to probe and understand.... Einstein and other leading scientists have stressed this difference, pointing out that beneath our scientific descriptions of the universe there may well lie 'realities' that we do not... comprehend"*[20] Thus the 'laws of science' are man's best attempts to describe the 'natural law' which we observe at work in our universe.

Isaac Newton revolutionised physics and cosmology when he published his universal laws of motion but these laws only *describe* the behaviour of the Universe. They do not explain *how* the laws of the Universe formed or what makes the laws work. In information terms, the laws describe the relationships between the basic data sets of the Universe (the properties listed in Table 2.2) which underlie its formation but they do not explain the information itself. To find that we need to seek the information that specified the 'Big Bang' which must have been a pre-requisite for the formation of the Universe and must have been generated prior to the Cosmos coming into existence. Paul Davies wrote: *"There is for me powerful evidence that there is something going on behind it all...It seems as though somebody has fine-tuned nature's numbers to make the Universe."*[21]

This "fine-tuning" of nature's numbers is completely unexpected if the Universe is simply a random, unplanned event. Andrews observed that the laws that describe the physical cosmos are almost universally expressed in terms of mathematics, such that the whole of nature appears to be built out of 'mathematical Lego'![22] Yet very few writers offer any explanation for the ultimate origin of natural law. They simply accept it as 'given' - as the unexplained and probably inexplicable substratum of reality that science seeks to explore and without which it could not exist.[23] In his book: *God; the failed hypothesis: How science shows that God does not exist,* Victor Stenger does attempt to answer this question.

Stenger wrote: *"So where did the laws of nature come from? They came from nothing!"*[24]

Andrews disagrees, explaining that although the flexible space-time constituent of the cosmos (its fabric) contains an awful lot of empty space, sometimes called 'the vacuum of outer space' it is not completely empty. He dubs this empty space 'void-one' to distinguish it from 'void-zero', the nothingness prior to the formation of the Universe.[25] However, the term 'vacuum' infers that the relevant space is 'empty'. Empty of what? Presumably mass and energy.[26] But Andrews point out that there is more to void-one than meets the eye. This space undergoes quantum fluctuations, as described by Heisenberg's uncertainty principle, that lead to the appearance and disappearance of phantom (or 'virtual') particles, which appear as particle-antiparticle pairs then immediately vanish again by mutual annihilation. More than that, void-one is filled with an 'ocean' of Higgs particles. This 'Higgs field' is a kind of cosmic treacle, resisting the movement of known particles and imparting to them the property of mass. But all this relates to empty space within the space-time fabric of the physical universe - a fabric that has, by definition, existed only from the moment of creation. Andrews concludes: *"the laws of nature follow naturally from the existence of the cosmos. Put more simply, they are just part of the created physical order."*[25] Thus, what we consider to be a 'vacuum' is only empty of *physical matter* but not necessarily empty of non-material entities, such as the laws of the Universe and information. So how did all this come about?

The Self-Organising Universe

The very existence of the Universe is a mystery. One proposal is that it assembled itself. Peter Atkins writes: *"Space-time generates its own dust in the process of its own self-assembly."* He named this process the *"cosmic bootstrap"*[27] but where did the information required for the assembly instructions originate? Another viewpoint on the formation of the Universe comes from Arabic teaching which has brought us the Cosmological Argument which states:

1. Whatever *begins to exist* has a cause;

2. The Universe had a beginning;
3. Therefore the Universe had a cause.[28]

This is intuitively correct. If I have a computer which I know has not existed forever and therefore had a beginning, then it must have had a cause, which is, that someone manufactured it. However, not all scientists agree on the causation of the Universe. Paul Davies writes: *"...for me it is much more inspiring to believe that a set of mathematical laws can be so clever as to bring all these things into being."*[29] The problem is that the information implicit in these laws had to come from somewhere or be generated somehow. Even if the laws 'did it', the information is still outstanding and demands an explanation. In effect, Davies' beginning of the Universe cannot have been the *ultimate* beginning because the information that defined our infant Universe also must have had a beginning; so how was that information generated?

Richard Dawkins' view is: *"The universe is nothing but a collection of atoms in motion, human beings are simply machines for propagating DNA, and propagation of DNA is a self-sustaining process. It is every living object's sole reason for living."*[30] However, this statement overlooks the information stored within the atoms, the humans and the DNA which dictates what they do. Inorganic processes do not create complex, organised systems without informational input.[31] Dawkins maintains that the Universe exists through the combination of chance and necessity. Chance implies that there is no information, whereas necessity requires that some form of intrinsic information is compelling a specific outcome to occur. But what is the origin of the information specifying the laws that created this necessity?

This question regarding the origin of the Universe is a distinct problem for materialists. Stephen Hawking and James Hartle developed a mathematical model to describe the beginning of the Universe and theorised that it came into existence through energy fluctuations in a quantum vacuum.[32] John Lennox explains that the expression *'quantum vacuum'* can be misleading as the word 'vacuum' tends to convey the idea that nothing is there at all. A quantum vacuum is a term that physicists use for an entity in its lowest energy state. It is certainly not *'nothing'*.[33] Professor

William Lane Craig points out that subatomic particles such as virtual particles do not come out of nothing. The quantum vacuum is a sea of fluctuating energy, an arena of violent activity that has a rich physical structure and can be described by physical laws.[34] There is another problem with Hawking and Hartle's solution too, regarding the *'spontaneous'* appearance of virtual particles that pop out of an empty vacuum. The rules of quantum mechanics, as expressed within the Heisenberg uncertainty principle, require that energy can be borrowed from the vacuum provided it is paid back very quickly. The more energy that is borrowed, the quicker the debt must be repaid.[35] So people who cite that a fluctuation in the quantum vacuum created the Universe are suggesting that the cancellation of the matter-antimatter particles went wrong and, defying quantum mechanics, the entire energy of our universe was extracted from this *'sea of fluctuating energy'*. But even if this was possible and all space, mass, energy and time were created this way, there is still no explanation as to where the vast amount of information came from to specify the workings of the Universe, unless there is information embedded in the energy fluctuations. However we do not know of any way that this could occur.

Thus a self-organising universe is beyond scientific explanation to date. This fact was colourfully expressed by Fred Hoyle in 1981, when he summarised the complexity of the Universe this way: *"Would you not say to yourself, 'Some super-calculating intellect must have designed the properties of the carbon atom, otherwise the chance of my finding such an atom through the blind forces of nature would be utterly minuscule.' Of course you would...A common sense interpretation of the facts suggests that a superintellect has monkeyed with physics, as well as with chemistry and biology, and that there are no blind forces worth speaking about in nature. The numbers one calculates from the facts seem to me so overwhelming as to put this conclusion almost beyond question."*[36]

So rather than decreeing that the Universe and its laws arose out of nothing, a more plausible supposition is that pre-defined information was introduced into the infant universe, at its formation, to sculpt the Cosmos.

The Multiverse

However, pre-specified information could not have occurred by chance, so David Deutsch of Oxford has introduced the concept of *'The Multiverse'* to describe the possibility of multiple parallel universes. It is proposed that as unlikely as it was for our universe to form the way it has, we may just be living in the infinitesimally rare universe that has all the parameters that, by chance, provided precisely the requirements necessary for life to exist. But despite multiple models for possible multiverses,[37] they all would require laws to govern them and bring them into existence. As described in Chapter 1, laws contain intrinsic information, which specifies them and that information had to come from somewhere. So who or what specified the information that describes the laws that control the universes within the Multiverse?

Andre Linde of Stamford suggested that we are living in an eternal superspace that caused the birth of multiple soap-like 'bubbles', each of which contains a different universe and ours is just one of them. Each bubble is finite in that it had a beginning and will have an end but the surrounding superspace is infinite.[38] The Multiverse option is also the model favoured by Hawking and Mlodinow in: *The Grand Design*. They conclude: *"Bodies such as stars or black holes cannot just appear out of nothing. But a whole universe can..."*[39] The problem is that Hawking and Mlodinow have failed to undertake an information audit. It may be possible in their theory for whole universes to appear out of nothing but it is impossible for meaningful information to do so. Each new universe without the appropriate laws and fine-tuning would be untenable for life.

A Self-Generating Universe

So, as an awareness of the significance of information within the Universe has grown, scientists have been exploring possible origins for the information and laws. In 2007, Paul Davies' considered the Multiverse option writing: *"This 'multiverse' explanation is just a little too glib. For example, what about the multiplicity of other universes? Not only is it extravagant to invoke all that cosmic real estate, there has to be some sort of universe-generating mechanism to create the varied patches in the cosmic quilt.*

Several mathematical theories exist that describe the creation of universes. In fact, one of these, called eternal inflation, is the currently favoured explanation for the Big Bang. ...However, the need to assume - without any explanation whatsoever - a universe-generating mechanism, an inflating space, and a host of other advanced physical ideas, leaves the multiverse theory falling far short of a complete explanation for why the universe is as it is."[40]

Alternatively, Davies suggests that the laws were initially poorly defined and they somehow became more precise as the Universe grew. However, there is not a shred of empirical evidence to support the notion that there is a materialistic explanation within our universe to account for the creation or refinement of itself or its laws. As with the 'cosmic bootstrap', information audit demonstrates that the laws had to predate the Universe or have been a product of the process that formed it. Thus the Laws of the Universe are the *product* of information not the refiner of it.

Hawking and Mlodinow also assert that the Universe generated itself from its own defining laws, which are the natural laws of the Universe. Unfortunately, the very laws to which they appeal for self-organisation must first themselves be specified by information![41] In a system which increases in complexity, there must be an introduction of information which must come from somewhere. Professor Bernd-Olaf Küppers explains: "... *Information conveying meaning cannot be compressed without change in, or even loss of, its meaning... This is the case for any meaningful information... Viewed technically, this means that no algorithms – that is, computer programs – exist that can extrapolate arbitrarily chosen parts of the message and thus generate the rest of the message. But if there are no meaning-generating algorithms, then no information can arise de novo. Therefore, to understand a piece of information of a certain complexity, one always requires background information that is at least of the same complexity. This is the sort-after answer to the question of how much information is needed to understand some other information.*"[42] Therefore there are no machines that can generate meaningful information out of nothing and the formation of the Universe with its complex structure of laws, stars and galaxies required information input. *Theoretically* this information might have been either intrinsic or extrinsic, however, information cannot be intrinsic to something that is not there yet, so it must have existed prior to its formation. Further, from Küppers statement, that information source must have been at least as information-rich as the information that is found in the Universe today!

The Physical Information Generator

So, whether you subscribe to one universe or a multiverse with an infinite number of constituent universes, this immense quantity of precisely specified information had to be generated somehow. A materialistic origin for this information still evades investigators but it cannot have appeared from nowhere, so it must have been generated by some mechanism and for simplicity I shall call it the 'Physical Information Generator' (See Figure 2.2).

The next problem is to account for this information generator.

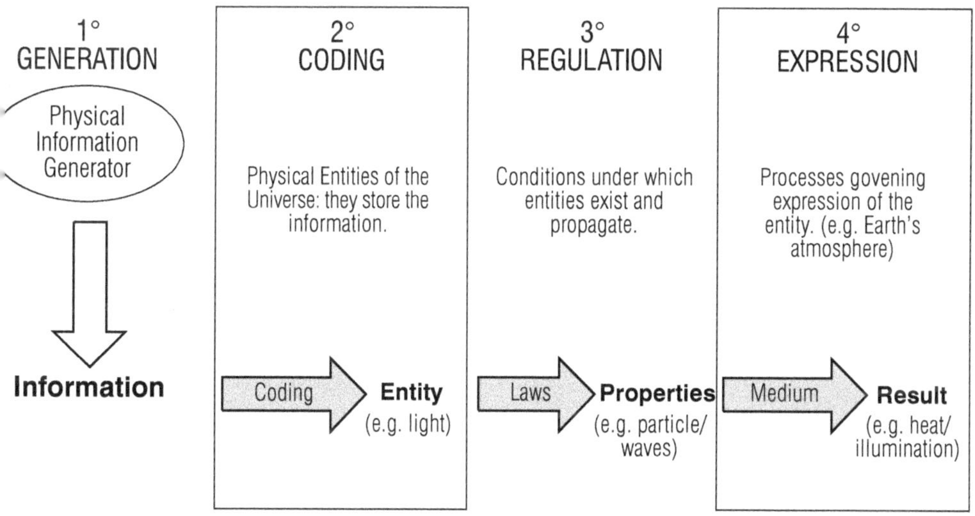

Figure 2.2: The Physical Information Generator
Arrows show the information flows. Notice that it is a one-way process. The illumination provided by the light that has been guided by the laws of the Universe cannot change the information that specified its composition or behaviour. An information generator is required.

Edgar Andrews points out that the origin of the universe: *"...can be modelled in various ways such as a 'hot big bang'...,but all such models point to an origin in which space, time, matter and energy came into being ex nihilo (out of nothing)...Current scientific cosmologies thus imply the existence of a non-physical realm (we'll call it 'eternity') that transcends space and time and within which the physical universe was created."*[43] This space was previously identified as 'void zero', outside our known universe. We do not have much information about this extra-universal space or the information generator yet but we will return to them in Chapter 7 after we have completed the rest of our information survey of the Universe.

Mathematical Universe

Since Pythagoras and Plato, scientists have wondered at the simplicity and predictability of the laws of the Universe and the role that mathematics plays.[44] In 1967, Eugene Wigner, wrote of the *"unreasonable effectiveness of mathematics in natural science."*[45] Fifty years later, Clara Moskowitz, Associate Editor of *Scientific American* introduced the term: 'quantum information' when she wrote in 2016: *"Lately scientists have begun to question...conventional thinking and speculate that space - and its extension according to general relativity, spacetime - is actually composed of tiny chunks of information. These chunks might interact to create spacetime and give rise to its properties, such as the concept that curvature in spacetime causes gravity."*[46] This could provide a basis for Wigner's observation. Meanwhile, Professor Leonard Susskind of Stanford states: *"It is not only that it becomes easier to describe* [the Universe] *with mathematics, as you go deeper and deeper into reality, mathematics becomes the only way to describe reality."*[47] Cosmologist Max Tegmark, sums it up by saying: *"To me, math is a window on our Universe... the master key to understanding what is out there. ...Equations... describe exactly what is going on...in our Universe...I think our Universe isn't just* described *by math, I think it* is *math. I think our entire Universe is a giant mathematical structure that we are part of. And that's the reason why the more we study physics, the more mathematical regularities we keep discovering...I think the*

Universe is a mathematical object...Space and time exist inside of it."[47]

Could the Earth and Universe have Evolved?

The question as to whether the Universe evolved depends upon which definition of 'evolution' is used: yes, it has changed but no, it did not create itself from nothing. The formation of simple structures such as rocky asteroids or a comet made of ices which might be specified by limited amounts of information may be envisaged but the spontaneous development of the myriad precisely defined and balanced constants, plus the laws of the Universe, is impossible because the generation or existence of information is a pre-requisite and our physical universe lacks an internal information generating mechanism.

There are also significant observations suggesting that 96% of the matter and energy in the Universe is 'missing'. So, for the Universe to be stable under the current models, it has been hypothesised that there is (yet to be discovered) 'dark matter' and 'dark energy' out there somewhere that prevents galaxies flying apart.[48] Our understanding still has a long way to go!

Mathematics: the fundamental language

So, our Universe is indeed an extraordinary entity! Paul Davies has stated that the laws of nature are real, objective truths about the universe that are found to be mathematical in form and we discover them rather than invent them.[49] Hawking and Mlodinow also point out that the Universe is extremely fine-tuned, and very little in physical law can be altered without destroying the possibility of the development of life as we know it.[50] Yet the explanation for this finely balanced universe had been observed by Galileo four centuries ago: *"The great book of Nature can be read only by those who know the language in which it is written, and this language is mathematics."*[51]

Perhaps most startlingly, in 1931 Paul Dirac predicted the existence of a previously unknown particle, the positron, using pure mathematics, to which Oxford Professor of Physics, Frank Close commented: *"That to me is one of the great profound mysteries of science,...It was a case of mathematics knowing about the Universe before we did.*"[52] But why should a randomly created universe exhibit precision and mathematical order at all, let alone foreknowledge of its own composition? The Standard Model offers no explanation.

Chapter 3

Is there Meaning in the Universe?

Lord Rees, President of the Royal Society, reflected on the extraordinary precision of six mathematical constants that define the form of the Universe and suggested that if any of them was changed even slightly, the Universe and the human race could not exist.[1] This implies that the likelihood that our universe occurred by chance is very small yet all the constants do fit together so that human beings, our biosphere and our universe form one integrated complex entity. How can such complexity exist, balancing on a knife's edge as it does? Normally complicated things that are unplanned end up as a jumbled mess. Why does the immense complexity of the Universe work? It is because information that synchronises harmoniously is embedded in the laws and fabric of the Universe, which allows the many inter-meshing systems to work like the cogs of the proverbial Swiss watch (pre-microchips of course). However, to properly understand our universe and hence discern true meaning we need to accurately perceive that information.

In Europe, the configuration of the Solar System was misunderstood until the invention of the telescope provided new data that allowed Copernicus, Galileo and Kepler to more accurately describe the motions of the planets, which subsequently allowed Newton to formulate universal laws. This permitted the apparent randomness of the movements of the planets to be explained logically and mathematically. The inaccuracies within the earlier

Greek model of planetary motion had disguised these mathematical laws. In the absence of obvious patterns, we often assume that there are no laws. However, a lot of what appears to be random or inexplicable within our Universe may yet be found to have structure and order but as yet, we have not found the appropriate 'telescope' to allow the accurate data to be collated and thus the laws to be revealed. One recent discovery is the Anthropic Principle.

The Anthropic Principle was first described by the Cambridge astrophysicist Brandon Carter in 1973, offering an explanation for Lord Rees' six numbers. It says that the seemingly arbitrary and unrelated constants in physics have one strange thing in common - they are precisely the values that are required to produce a universe capable of sustaining life and in particular, human life.[2] The Universe gives the appearance that it was designed or finely tuned to support life on Earth as, apart from the six constants highlighted by Lord Rees, physics textbooks are littered with precisely defined relationships that are necessary for our universe to sustain life.

In 2008, *New Scientist* published an article by Stuart Clarke entitled: *"Unknown Earth: Our Planet's Seven Biggest Mysteries."* In this article Clarke asks the question: *"How come earth got all the good stuff?...We know that its distance from the Sun provides the right amount of heat and the light to make the planet habitable...Without the unique mix of carbon, hydrogen, nitrogen, oxygen, phosphorus and sulphur that makes up living things, and without liquid water, life as we know it could not have evolved."*[3]

The fact that we are here and can observe the Universe, implies that the fundamental constants *must* be 'just right' to produce life. However, this is circular reasoning, because if the constants were not 'just right', we would not be here to comment. Even so, the brutal facts are that the Universe does *not* seem to be a random or chance event. We can postulate a multiple universe scenario, in which only one or some universes produce life, but we cannot validate that scientifically because we can only investigate the *one* universe in which we live. So, let's dig deeper.

Extraordinary Carbon

Carbon is unique among the elements in its ability to form long chains with itself. As one of the lighter elements, its atomic volume is relatively small, allowing it to fold into many different configurations. Other elements, such as silicon, have similar bonding characteristics but the sizes of their nuclei prevent the formation of the large folded chains that carbon does, to produce the vast array of uniquely shaped proteins. This is why most scientists conclude that all life in the Cosmos must be based on carbon.

However, the synthesis of carbon depends upon a remarkable coincidence in the properties of the fundamental forces in the Universe. In the 'Big Bang' Model, plenty of hydrogen and helium were produced but no carbon. So where did the carbon in living organisms come from? In the 1950s it was realised that most of the chemical elements heavier than helium were manufactured in the cores of stars, as the product of nuclear fusion reactions; the energy released by these reactions makes the Sun and stars shine.

However, although most elements can be made by collisions between two smaller atomic nuclei, carbon cannot be made this way because all the intermediates from helium to carbon are highly unstable. Cambridge astronomer Fred Hoyle deduced that carbon might form from the simultaneous collision of three helium nuclei but the probability was miniscule unless some other factor was influencing the result. Hoyle knew that if the masses and energies of the colliding particles that go to form an element are just right, the chance of successful reactions can sometimes be greatly amplified by the phenomenon of resonance, similar to the way that an opera singer can shatter a glass by hitting a certain note. Carbon just happens to be one of these.

So, what determines the carbon resonance? Ultimately it depends on the strength of the force that binds protons and neutrons together in the nucleus, which is one of the four fundamental forces discussed in Chapter 2. If the strength of this force was fractionally stronger or weaker, human beings would not be here to know.

Physicist, Paul Davies wrote: *"Hoyle himself was deeply impressed by this discovery. 'It looks like a put-up job,'....A similar conclusion was reached by the Princeton physicist Freeman Dyson:* [who stated] *'In some sense, the universe knew we were coming.'"*[4]

Amazing Water

Water complements carbon perfectly. However, the very nature of water, which is so vital to life, is something of a mystery too, a point noted in 1913 by Harvard chemist Lawrence Henderson, one of the forerunners of anthropic reasoning.[5] Unique amongst the molecules, water is less dense in its solid state than its liquid form - hence ice floats! If it did not, the oceans would freeze from the bottom up and Earth would now be covered with solid ice. This property derives from the unique properties of the hydrogen atom. However the properties of water are best showcased when they are studied in Earth's biosphere. Life relies on water being liquid. Water must be in its aqueous phase for molecules to be able to move and react (unlike ice), but dense enough for reactions to take place sufficiently often, that is, fast enough, to support life (unlike steam). It has been estimated that if the Earth was five per cent closer to the Sun there would be a runaway greenhouse effect but if the Earth was twenty per cent further away from the Sun then carbon dioxide would condense in the atmosphere and our planet would freeze.[5] The Earth is fortunately the perfect distance from our star, the Sun, for water to remain in its liquid phase all year round. Hawking and Mlodinow write: *"...given any star, scientists define the habitable zone as the narrow region around the star in which temperatures are such that liquid water can exist. The habitable zone is sometimes called the 'Goldilocks zone', because the requirement that liquid water exist means that, like Goldilocks, the development of intelligent life requires that planetary temperatures be 'just right'."*[6]

Without the precisely configured environment that is found on Earth, the water cycle cannot operate and life could not exist as we know it. Due to these multiple constraints, it may not be possible for life to exist anywhere other than Earth, unless almost identical conditions to those found in our solar system are replicated elsewhere in the Universe.

So why do carbon and water behave the way they do? It is their intrinsic chemical properties, which include their melting points and vaporisation points, which are controlled by their molecular structures. These, in turn, depend upon the arrangement of their atoms, which derive from their proton and electron configuration.

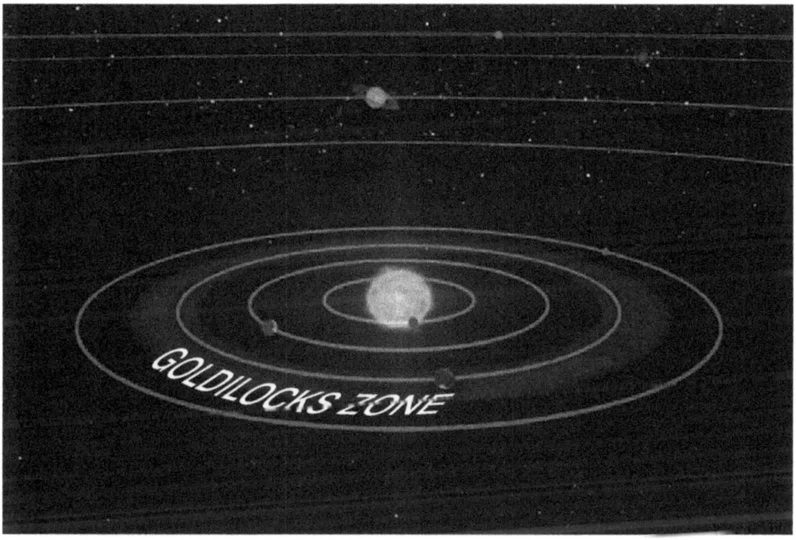

Figure 3.1: The Circumstellar Habitable Zone or 'Goldilocks Zone'[7]

Tracing the chain back to its simplest form, their sub-atomic structures depend upon the forces and laws that govern the sub-atomic particles within the Universe, which appear to have been active since the beginning of the Universe. These laws are determined by intrinsic information, which must have been in existence for at least the same period as the laws.

Fine Tuning

So, our material universe is underpinned or specified by intrinsic information. But the *most* unusual feature is not that the information is there, but that it weaves together so perfectly to make life possible. Over one hundred criteria have been described that require tightly controlled values for life to exist on the Earth.[8] Table 2.2 listed some of the sub-atomic characteristics required for a stable universe. Table 3.1 lists a few of the properties of the Earth and the Cosmos necessary for life. The precision of this balancing act is illustrated by astrophysicist Hugh Ross who observes that unless the number of electrons is equivalent to the number of protons to an accuracy of one part in 10^{37}, or better, electromagnetic forces in the universe would have overwhelmed

	Properties	Effects if too great	Effects if too small
1	Earth's distance from the sun	If 20% greater, all water freezes[11]	If 5% less, runaway greenhouse effect[11]
2	Size of Earth (surface gravity)	Too much methane & ammonia[11] Comet impacts destroy life	Water & oxygen lost; molt core not protective of radiation[11]
3	Earth's rotation speed (Length of a day)	Winds velocities become destructive[12]	Earth's surface gets too h by day and too cold by night[12]
4	Orbital eccentricity (climactic stability)	Temperature and climate vary too much for life[13]	Nil. Orbit is almost circula creating a stable climate[13]
5	Earth's axis tilt - 23½°	Seasons are too disruptive due to temperature variation[14]	No seasons or monsoon causes aridity (less arable land) [14]
6	Earth's distance from the moon	Insufficient stabilisation of Earth's orbit[12]	Ocean tides become too disruptive[12]
7	Size of the Moon (large at ¼ of Earth)	Tides become disruptive[15]	Loss of stabilisation of Earth's rotation[15]
8	Size of the Sun (governs luminosity)	Gravity prevents Earth's rotation, making it too hot for life[13]	Goldilocks Zone is too shallow for eccentricity in Earth's orbit[13]
9	Wavelength of the sun's light	High energy wave-lengths damage life (e.g. skin cancers)[13]	Low energy wave-lengths are insufficient for photosynthesis[13]
10	Variability of the Sun	Risks sterilising events on Earth[13]	Nil. Constancy is good[13]

Table 3.1: Properties of the Universe[10]

gravitational forces so that galaxies, stars and planets would never have formed.[9]

So, life depends upon the precise definition of both the properties of the large objects in our universe as well as the fine structure. But many people will be surprised to discover that, as small as these fine structures are, they are mostly vacant 'space'. In the 1920's, electrons, protons and neutrons were thought to be hard particles but it is now known that there are vast empty spaces between atoms that are a thousand times greater than the space it takes to contain an atom thus, so-called 'solid matter' is mostly empty space. Were it not for the rapid motions of the molecules, matter would readily interpenetrate and 'solid' objects would not exist. This would mean we would have trouble standing, sitting or supporting anything at all.[16] If the atom is envisaged as an Olympic sports stadium and a grain of sand placed in the centre of the stadium's arena represents the proton, then the electron would be a gnat on the upper, outermost terrace for spectators and when one does rarely encounter an atom, it is not composed of hard particles but electrically charged waves. Since energy and matter are equivalent, one can say that solid matter is the way that space happens to be vibrating at a certain time and place.[16] Thus, 'material' objects are not as 'solid' as they initially appear, yet the intricacy is profound!

The sheer weight of evidence regarding fine tuning has forced some scientists to rethink their response to the more than one hundred cosmological coincidences that have been identified. Initially they were attributed to fortuitous accidents, as evidenced by an early quotation from Hoyle (1951): *"I have often seen it stated that our situation on earth is providential. The argument goes like this. It is providential that the earth is of the right size and at the right distance from the sun. It is providential that the sun radiates the right kind of heat and light. It is providential that the right chemical substances occur on earth. A long list of this sort of statement could be compiled, and to some people it looks as if there is indeed something very strange and odd about our particular home in the universe. But I think that this outlook arises from a misunderstanding of the situation. Because if everything was not just right we should not be here. We should be somewhere else (or more likely, nowhere)."* However, thirty years later, in light of all the intervening discoveries, Hoyle wrote: *"Such properties seem to run through the fabric of the natural world like a thread of happy coincidences. But there are so many odd coincidences essential to life that some explanation seems required to account for them." (1983)*[17]

The Tangible Domain

Critics of the fine-tuning theory of the Universe have described the Weak Anthropic Principle which states (as Hoyle suggested in 1951) that if the universe were not fine-tuned for life, then human beings wouldn't be around to observe it.[18] This argument proposes that the Universe as we know it today was necessary for the existence of life but design was not required. Advocates of this theory often suggest the existence of a multiverse, with countless other universes that do not support life; we just happen to live in this one that does. There is no way of detecting these other universes however, so there is no scientific way to prove or disprove this hypothesis. Thus, by the conventions of science, the Multiverse is not 'scientific' but 'philosophical'. However, if we once again resort to looking from an information vantage point, the Universe is undoubtedly filled with meaningful information, which had to be specified at some point; which is to say that it required an information generator rather than being the product of random chance.

Most of the information discussed in this chapter relates to material objects that enable our physical existence and they exist within a touchable or *Tangible Domain*. The relationships between the entities specified by this information can be mapped as shown in Figure 3.2.

The way that the pieces mesh together is extraordinary which provoked Paul Davies to write: *"It seems as though someone has fine-tuned nature's numbers to make the universe...The impression of design is overwhelming."*[19] So our Universe, as evidenced by the charge on an electron, organic chemistry, DNA, life, the Earth

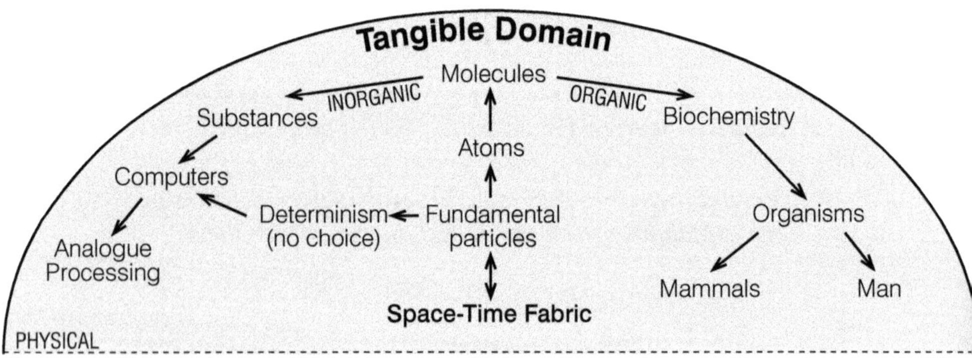

Figure 3.2: Information Map of the Tangible Domain
This outline of some of the core physical entities in the Universe shows how information is inter-related and the way these relationships can be mapped.

and our solar system among many other examples is intricately harmonised and has the appearance of design. The question is: why should physics, chemistry, cosmology and biology all be finely tuned and exhibit extreme specificity, when we are taught at school and university that the Universe occurred by chance; it happened 'by accident'?

When describing the origin of life in *Scientific American* in 1953, George Wald wrote: *"Time is in fact the hero of the plot... given so much time the 'impossible' becomes possible, the possible probable and the probable virtually certain. One has only to wait: time itself performs miracles."*[20] The problem with this explanation is that the forces and laws of the Cosmos will not make a stable universe or biosphere without first possessing precisely specified meaningful information, even with billions of years at their disposal. Without meaningful information the Universe would be a bland, sterile, empty place. Thus, the true hero is invisible, intangible information!

Chapter 4

Searching for the Meaning of Life

Finding the meaning of life has been the Holy Grail of philosophy for millennia but where to start? Well, we previously observed that meaning in the universe was imparted through information, so what can information tell us about life? Living organisms are saturated with information to the extent that Bernd-Olaf Küppers wrote: *"Ever since the elucidation of the molecular basis of living systems, we have known that all elementary processes of life are governed by information. Thus, information turns out to be a key concept in understanding living matter."*[1] So, going back to these *"elementary processes of life"*, is there meaning to be found in the information that was required for lifeless atoms and inorganic molecules to become the first living cell – when life first appeared?

The Law of Cells (Pasteur's Hypothesis)

In the early Victorian era, the concept of *vitality* (the spontaneous generation of life) was a widely held belief because, as microbes could not be seen, it appeared that illnesses just spontaneously occurred causing people to sicken and die. Maggots appeared in dead meat while hay spontaneously grew mould and rotted, so simple life forms appeared to be able to spontaneously form 'out of thin air'. The explanation given for these observations was that it was caused by *'miasma'* ('bad air'). Perhaps it was this widely held concept that led Charles Darwin to write to his friend, Joseph Hooker in 1871, introducing the idea of life spontaneously forming in a *"warm little pond"*. Darwin may not have been aware that only nine years earlier, in 1862, the French chemist and microbiologist, Louis Pasteur showed that spontaneous generation did *not* happen.

Pasteur's findings have subsequently been formalised as the 'Law of Biogenesis' which states: *"Life comes from life"* which became the first Law of Biology. However in 1870, Thomas Henry Huxley, a colleague and supporter of Charles Darwin and a champion of the Theory of Evolution, coined the term *abiogenesis*. In an essay he wrote: *"Carbon, hydrogen, oxygen, and nitrogen are all lifeless bodies. Of these, carbon and oxygen unite in certain proportions and under certain conditions, to give rise to carbonic acid; hydrogen and oxygen produce water; nitrogen and other elements give rise to nitrogenous salts. These new compounds, like the elementary bodies of which they are composed, are lifeless but when they are brought together, under certain conditions, they give rise to the still more complex body, protoplasm, and this protoplasm exhibits the phenomena of life. I see no break in this series of steps in molecular complication..."*[2] Huxley was partially correct, in that simple elements are lifeless and protoplasm supports life, but he grossly underestimated the complexity of the process by which simple elements link together to form living protoplasm, which we now call *cytoplasm*. Huxley and Darwin lacked the electron microscopes and modern investigative techniques we have today which have revealed the information-dependant nano-machinery necessary for all cells to function.

If Huxley was right and life did begin from basic components, then all the essential ingredients needed to be assembled to commence functioning as a cell. Yet, this has not been achieved even in a laboratory, so the chance of it happening in a warm little pond is even less probable to the point of being impossible. American biologist, Craig Venter has *copied* a bacterial genome and then inserted it into an already living cell cytoplasm but this is vastly different from designing, as well as constructing, the bacterial code, the cell membrane, mitochondria and all of the cytoplasmic elements from scratch and then finding that the cell is now alive.[3] The Neo-Darwinist will reply that such complexity need not have existed at the start and that, as Darwin *hypothesised*, a simple model can grow through the process of evolution to achieve a very complex result. However, Pasteur *demonstrated* that without a pre-existing working model, a new cell does not begin: *a living cell is required to produce another living cell.*

The corollary of Pasteur's Law is that once a cell dies it cannot be reanimated unless it begins again from both its *component DNA* (information) and another *living cell.* To test this, try growing a cell without its nucleus (which contains the DNA). Alternatively, observe that a dead cell which still contains the correct DNA will not grow either. It is not just any DNA that will do, however; both the quality and quantity of the DNA must be exquisitely specified. This precision suggests that the information required to produce another living cell must be just as carefully controlled.

The sequencing of the human genome in 2000 revealed the vast amount of information required to specify a human being but it has also shown the efficiency with which this information is packaged. Not only does the human DNA sequence specify the materials out of which we are constructed but it provides the construction process so that the correct procedure can be followed. The microscopic machines that carry out the construction process are inherited from the cytoplasm of the living parental cell from which the new organism has grown. This is called cytoplasmic inheritance and the DNA of the cytoplasmic nano-machines is *not* coded in the DNA within the nucleus.

DNA and Information

Richard Dawkins wrote: *"What lies at the heart of every living thing is not a fire, warm breath, nor a 'spark of life'. It is information, words, instructions...Think of 1 billion discrete digital characters...If you want to understand life think about digital technology."*[4] Living cells and computers process information in similar ways. Computers possess 'hardware' and 'software' and living cells do too. The 'hardware' in living organisms is called the body and the information that specifies it is stored in both the DNA in the nucleus and the nano-machines in the cytoplasm. There is also epigenetic information, which is *metadata*, the information that controls the other information. It runs cell processes as well as some other automated systems such as blood pressure control.

These processes are hidden within all living organisms and are largely invisible until you actively search for them. This cellular software information that is stored in DNA grows our bodies from a single cell and then run all our functions.

We take DNA for granted today as the ubiquitous store of information in all living things but it is a very powerful yet very simple language that spells out the web of life. It has an alphabet (A, C, G, & T) and grammar (rules of engagement: A=T; C=G). These letters are symbols for the nucleobases: adenine (A), cytosine (C), guanine (G) and thymine (T). Its power resides in its ability to store immense meaning. However, meaning in spoken or written language does not happen by accident and the meaning encoded in the genetic language of DNA could not have been placed there by human beings, so where did it come from? For millennia, people had believed the explanation given in the Bible which states that God created life but Charles Darwin noticed similarities between living organisms and fossil specimens and proposed the Theory of Evolution.

Defining Evolution

Charles Darwin was not the first person to describe the concept of evolution but he was the first to provide a naturalistic mechanism as to how it might have occurred. There are various ways to define 'evolution' and they are not the same. To avoid confusion, they need to be precisely understood. The first is a broad definition stating: *"The appearance in orderly succession of a long train of events."*[5] Fowler and Kuebler comment: *"This is what one could call the basic or 'wide' meaning of evolution; it corresponds to what is least controversial about evolution, the simple notion of change over time."*[6] The other narrower definition is expressed as: *"Origination of species of animals and plants by a naturalistic process of development from earlier forms, and not by a process of 'special creation'"*.[5] Darwin used the term: *"Descent with modification"*.[7]

More recently, the terms 'macro' and 'microevolution' have appeared. Macroevolution describes the process where Darwin hypothesized that the information in one species can increase to the extent that it changes *into* another species while microevolution describes the fluctuations of information *within* a species. A species is a group of organisms that can interbreed to produce fertile offspring.

If evolution is defined as: *"random variation with natural selection"* as Darwin did, then any change, including the *loss* of information, may be viewed as 'evolution'. But from an information perspective, we perceive that change, without the *addition* of information, will not build more complex organisms.[8] Change alone (microevolution), cannot explain the vast diversity of life on Earth. To change a single-celled amoeba into a mammal or a man (macroevolution) requires the addition of *new* information with increases in both its quantity and its complexity. Once we delve into the realm of information relating to evolution, there is a need to distinguish between the appearance of *new traits* and the formation of *new information* because the rearrangement of existing information code can create a new trait *without* any new information. A typical example is the development of bacterial resistance to drugs by the shutting down of cell pumps. This is important because rearranging deck chairs on a cruise ship doesn't build a bigger, faster ocean liner.

The Neo-Darwinian Synthesis[9]

Human beings were not present to observe and record macroevolution happening in real time, so macroevolution cannot be proven by the first method of science (direct observation), to verify the progressive change that has been predicted. It has not been repeated either, so the second method of scientific investigation is also denied us. Therefore, scientists must resort to indirect means to establish a case, but as the evidence (such as the fossil record) is incomplete and evolution involves excessively slow processes over exceedingly long time periods there is inevitably a lot of inference and the result is not surprisingly controversial, as evidenced by the multiple disputes. Fowler and Kuebler (who conclude that macroevolution does occur) write: *"While there is solid evidence that these mechanisms can lead to small-scale change (Microevolution), a fact accepted by all schools* [of thought], *there is at present little direct evidence that they can explain* all *evolutionary change (Macroevolution); what evidence we have is primarily circumstantial....Neo-Darwinians argue that the available evidence supports their theory, and no other theory has a better explanation of this evidence, so they infer the truth of their theory by* inference to the best explanation. "[10] The evidence and counter arguments are summarised in Table 4.1 on the following page.

It is very important to avoid confusing observational data with mechanisms. The existence of fossils that appear similar cannot prove relatedness unless there is a 'written pedigree' to prove the relationship, as is required in pedigree breeding associations. Neo-Darwinists are hoping that molecular biology and DNA sequencing will provide this evidence but the data to date suggests the lineages are 'mixed up'.

If Darwinian Evolution is true, then once the first cell came into existence, no matter how simple it was, it must have developed a mechanism to reproduce before it died or all the information that it had acquired would have been lost. However, in living things today, reproduction is a rigorously controlled and very detailed process. Even in the simpler *asexual* reproduction, as in bacteria, special apparatus is needed to separate the contents of the daughter cells with enzymes to unzip the DNA for copying and they also would have had to evolve. The biggest problem here is that instead of millions of years to achieve this ability, the first cell must successfully accomplish this intricate and complex task within its own lifetime, which is totally outside the capability of the Neo-Darwinian model.

Mutations

Darwin thought that accidental mutations would cause species to continually change, yet most mechanisms in living organisms cause natural selection to move back towards the norm rather than evolve new forms to drift further apart as Darwin supposed. Factors in human beings and mammals include:

a. The extreme accuracy of replication;
b. Cell mechanisms repair most mistakes in transcription;
c. The immune system eliminates most damaged (altered) DNA;

	Evidence[11,12]	Argument	Counter argument
1	Geologic Column (layers within the Earth's crust)	Suggests the history of life on Earth is long enough for macroevolution to have occurred.	Organisms appear suddenly without enough time to acquire complex information.
2	Fossil record	Fossils will be found demonstrating the transitions of one life form into another.	Finely graded transitions are not found yet; it is circumstantial evidence; needs written pedigree.
3	Embryology	Embryos show the similarity of our evolutionary past.	Haeckel's drawings were grossly inaccurate and misleading.[13]
4	Random variation with Natural selection	Darwin's finches and peppered moths demonstrate adaptation of organisms to their environment.	There is agreement here. Most scientists agree that microevolution is true.
5	Bacterial resistance	Proves microevolution occurs and by extrapolation, macroevolution.	This is information loss not gain which is microevolutionary.
6	Selective breeding	If it works in captivity it demonstrates 'evolution in action'.	Selective breeding is not random and is microevolutionary.
7	Population genetics	Mathematical models demonstrate macroevolution over long timeframes.	Extrapolating microevolution into macroevolution is invalid; the models are non-biological.
8	Anatomical similarities	Demonstrate body plan relatedness showing descent from a common ancestor.	Exhibits similar information but not its origin; need a pedigree to prove universal common descent.
9	Molecular biology	Biochemical pathways will confirm the path taken by evolution.	Unproven to date; more data is needed to demonstrate linkages.
10	DNA (genetics)	Common genetic code will demonstrate the information pathway from simple to complex organisms.	There is a confused DNA tree to date which does not support universal common descent.
11	Geographic distributions of flora and fauna	Variations in species and subspecies around the world demonstrate gradual global evolution.	Demonstrates migration and microevolutionary survival of the fittest.[14]

Table 4.1: Summary of the Neo-Darwinian Synthesis

d. The new mutation must be selected as *the* gamete (cell) used for reproduction;
e. The new mutation must give the cell/organism an advantage *at every step* of its mutation pathway or it will be eliminated by natural selection.

Every new mutation must run the gauntlet of these five steps to become established in the genetic pool otherwise it will be eliminated by natural selection. This five step process has strong survival advantage as ninety-nine percent of mutations are harmful and without it, all organisms would become extinct within a few generations due to the accumulated mutation-load.[15] When the operating systems in our computers become corrupted, we reboot them and start again with the same machine but with living organisms, the result is death, so that model is discarded and the process needs to start again with another individual from the same species but with fewer defects.

British microbiologist, Alan Linton illustrated the absence of macroevolution when he wrote: *"Throughout 150 years of the science of bacteriology, there is no evidence that one species of bacteria has changed into another."*[16] Bacteria can replicate within hours so this time-period for bacteria corresponds to 2 to 3 million human-year equivalents, during which time Darwinists tell us there was enormous change in complex species. Richard Lenski's team at Michigan State University has cultured six generations of E. coli bacteria every day for over thirty years which has involved billions of bacteria and the equivalence of nearly a million human years of evolution,[17] yet no new species have emerged. Richard Dawkins highlighted the acquisition (after 33,000 generations) of the ability to metabolise citrate, which formed part of the culture medium of the experiment. This, he declared, was proof of evolution *"before our very eyes"*.[18] In reply, biochemist, Jonathan Sarfati observed that E. coli can already use citrate directly under anaerobic conditions, but this type of metabolism is less efficient than aerobic respiration, so it is normally switched off in the presence of oxygen. In this case, there was a loss of regulation allowing this anaerobic citrate pathway to operate all the time, which is less resourceful and informationally downhill.[19] The problem is that mutations only delete meaningful information or increase the random variability within the gene pool, which does not create novel information to build new assemblies. New structures require multiple complementary changes over many generations, which has *never* been observed.

In fact, the reverse is observed. Over many generations, the build-up of random mutations is increasingly likely to damage the cell's operating machinery, so there needs to be a mechanism to remove the damaging mutations before they compromise the specie's gene pool. Natural selection performs this function, to minimise mutations that may be harmful and in so doing, natural selection causes variation within species to return towards the mean, rather than away from it. James Shapiro, Professor of Biochemistry observes: *"It has been a surprise to learn how thoroughly cells protect themselves against precisely the kinds of accidental genetic change that, according to conventional [Darwinian] theory, are the sources of evolutionary variability. By virtue of their proofreading and repair systems, living cells are not the passive victims of the random forces of chemistry and physics. They devote large resources to suppressing random genetic variation..."*[20] So these successive layers of error correction show that life puts an exceedingly high value on maintaining the integrity of its information systems. This suggests that there is a need for extreme accuracy in the copying process, because if it was not required, it would not be preserved as it is obviously a complicated and hence 'expensive' process to maintain.

Biochemical (Molecular) Evolution

All biological systems can ultimately be traced back to their biochemical building blocks and we know that Darwinian Evolution can only move one small step at a time, so if Darwin's hypothesis is true, there must be a biochemical evolutionary pathway for every protein, every enzyme system and every organelle in every living creature, yet there is no detailed traceable trail demonstrating how evolution could have occurred at a molecular level for even one of them, even though the Neo-Darwinian Theory *requires* that they existed. Michael Behe, Professor of Biological Sciences writes: *"Molecular evolution is not based on scientific authority. There is no publication in the scientific literature - in prestigious journals, specialty journals, or books - that describes how molecular evolution of any real, complex, biochemical system either did occur, or even might have occurred. There are assertions that such evolution occurred, but absolutely none are supported by pertinent experiments or calculations...Despite comparing sequences and mathematical modelling, molecular evolution has never addressed the question of how complex structures came to be."*[21]

This is corroborated by James Shapiro who also acknowledges that there are no detailed Darwinian accounts for the evolution of any fundamental biochemical or cellular system: only wishful speculation.[22] While microbiologist, Michael Denton observes: *"In terms of their biochemistry, none of the species deemed 'intermediate', 'ancestral' or 'primitive' by generations of evolutionary biologists, and alluded to as evidence of sequence in nature, shows any sign of their supposed intermediate status."*[23] Thus there is no gradual evolutionary information progression demonstrated in molecular biology, which there must be if Darwin's theory is true.

So, if evolution does not generate new information for species change, what does it do? Biologist and philosopher Paul Erbrich explains that the mutation-selection mechanism is an optimisation device that enables an already existing living organism to adapt to changing environmental conditions, however it does not create anything radically new.[24]

Information in Biological Systems

Richard Dawkins writes: *"Biology is the study of complicated things...Physics books may be complicated, but...[t]he objects and phenomena that a physics book describes are simpler than a single cell in the body of its author."* He notes that each nucleus contains a digitally coded database, larger in information content than all 30 volumes of the Encyclopaedia Britannica put together, yet the author is composed of trillions of these cells.[25] Thus, our biological information system is ultra-compact and *extremely* large.

John Lennox gives us some appreciation of the inner workings of a cell when he writes: *"It is hard for us to get any kind of picture of the seething, dizzying complex activity that occurs inside a living cell, which contains within its lipid membrane maybe 100 million proteins of 20,000 different types and yet the whole cell is so tiny that a couple of hundred could be placed on the dot in this letter 'i'."*[26] Within the cell, where the workers are proteins, with so much happening in such a confined space, the activity is frenetic with the roads (cytoplasmic reticulum) being absorbed and rebuilt constantly as materials are moved around within the cell. Much of this activity is for protein production that is vital for cell survival.

Protein Synthesis

To produce new proteins, the command centre (the regulatory genome) in the manager's office where the blueprints are stored (nucleus of the cell) sends a message (RNA) to copy a section of the master blueprint (DNA) to create a messenger (mRNA). The mRNA is transported to the factory floor (cytoplasm) where the protein manufacturing machines (ribosomes) take the information and translate it into a protein (string of amino acid). Ribosomes are universal to all living organisms.[27] In 1999, Roger Garrett, Professor of Molecular Biology wrote: *"The ribosome... is probably the most sophisticated machine ever made. All of its components are active and moving..."*[28] In plant and animal cells, ribosomes contain 73 proteins with four ribosomal RNAs integrated into one unit. Even in simpler bacterial cells, the ribosomes are constructed from 50 proteins with three ribosomal RNAs.[29] As proteins cannot be made without them and every living cell requires them, ribosomes would appear to have been necessary for the survival of the first cell too. The conundrum is that as the ribosome is one of the most complex nano-machines ever discovered, how did the first cell acquire the information to build such complex machinery, which appears to have remained unchanged for the next 500 million years? This does not sound like Neo-Darwinian Evolution!

Information in Proteins

Protein molecules are both the building blocks and the workhorses of life and contain an immense amount of information. The average protein contains between 100 and 500 amino acids. There are over 150 naturally occurring amino acids but only 20 of them occur in living organisms. These can be arranged in any order. If a protein molecule was 100 amino acid units in length, there would be 10^{130} alternate ways to arrange the amino acids for that DNA string.[30] This figure is so large it is difficult to comprehend but to put it into perspective, physicists estimate that there are about 10^{80} particles in the Universe so 10^{130} is an extremely large number. Usually, there is only one arrangement of this vast number of alternatives that will provide an efficient biologically active molecule. To identify this one active combination of amino acids by chance is unimaginable unless there is a guiding mechanism. To complicate the situation further, the protein must be precisely folded to form a specific three-dimensional shape to make the protein molecule functional. Hulbert Yockey, Cambridge nuclear physicist and bio-informatician, calculated that if a supply of suitable amino acids had been available for the estimated period that life has been on the Earth of 1 billion years, the largest protein that would have been produced by chance would be just 49 amino acids long.[31] The likelihood of this protein being biologically active would be extremely remote. Thus, in real life situations, there is a vast amount of information but it does not readily form functional relationships by chance. Further, in life, only *meaningful* information can produce a useful outcome.

Energy Management

Cells must be active to survive so even the first organism needed an energy source. Photosynthesis is the usual mechanism for generating energy in biological systems, including plants, algae and cyanobacteria. The process of photosynthesis is intricate and depends upon an extremely complex 'catalytic core'[32] within the chloroplasts of green plants. The information required for the chloroplast complex is enormous and requires precision construction. As the core complex is universal to photosynthetic organisms, it does not appear to have evolved over time and, in fact, there is no evidence it evolved at all. Even the cyanobacteria, which are considered by many biologists to be 'very ancient' and hence presumed to be 'primitive', contain this mind-bogglingly complex nano-machine. This means that the information has been present throughout the entire history of green plants. Thus, after the initial generation of the information which built the first chloroplast, there has been information stasis.

The next problem is that ATP is the universal energy storage in *all* living things. It is manufactured by a highly efficient nano-machine called *'ATP-synthase'* which is found in mitochondria in animals and chloroplasts in plants. Since all life needs ATP-synthase, it must have formed with the first cell or very early after the formation of life but ATP-synthase is another extremely complex machine and is required for the cell's daily survival. So how could it have evolved so early in the history of life on Earth? It is significant to note that the first plants required both the photosynthesising catalytic core and ATP synthase!

Internal Cellular Transport

The intracellular environment is considered to have the viscosity of yoghurt,[33] so it would take a long time for chemicals to move around the cell if they had to get there by random diffusion. Edgar Meyhöfer, Associate-Professor of Biomedical Engineering describes how every living plant and animal cell has an internal highway system of microtubules on which a transport molecule or motor protein called kinesin travels. It is similar to the road network in your town and the kinesin is like a two-legged robotic mule that literally walks along the 'roads'. Unlike the roads that you are used to however, the microtubules are made as they are required and then reabsorbed and recycled. But instead of consuming hay, kinesin uses ATP for fuel (one ATP molecule per eight nanometre step).[34] The sophistication of this complexity was completely unforeseen only a generation ago yet once again the first cell would have required an internal transport mechanism to survive.

Complex Biochemistry and Catalysts

Organisms also depend on interlocking biochemical information systems. To delve into the intricate biochemistry of living cells would require several large textbooks but due to limited space we will look at enzymes as a typical example.

Each living cell is filled with many, many chemicals. When I was at school, I could never understand why all the chemicals in the cell did not inappropriately interact with each other, causing chemical catastrophe and cell death. However, it is now realised that most reactions in cells are so slow that, without the right enzyme, effectively, they don't occur at all. Yet with the right enzyme they happen very quickly. These enzymes or catalysts are special facilitator molecules, which may be a chemical or a protein that can speed up the chemical reaction between two other substances by a billion or even a trillion times[35] and yet they are not consumed in the process. Richard Wolfenden, Professor of Biochemistry wrote: *"Without catalysts, there would be no life at all, from microbes to humans...It makes you wonder how natural selection operated in such a way as to produce a protein that got off the ground as a primitive catalyst."*[36] And each reaction requires its own *unique* catalyst so large numbers of catalysts are needed but each in only tiny quantity.

In a normal cell, the supply of catalysts is tightly controlled, so that in the absence of the catalysts most cellular chemicals do not interact at all. Rather than suffering from runaway chemical calamity, the first cell must have sat around for literally aeons, waiting for its chemical reactions to occur. In practice, cells without catalysts would have died before the necessary minimal set of catalysts could have evolved. Richard Dawkins suggests that low efficiency catalysts would have sufficed to begin the first cellular reactions, before natural selection made the processes more efficient over time. However, dozens, if not hundreds of catalysts had to be present from the start, even in the most minimal cell and the lower the efficiency of the initial catalysts, the more important it would have been to have *all the catalysts present* as reactions would have been slower. This necessity for the simultaneous development of multiple chemical accelerators in parallel with the chemical pathways themselves, even in the first cell, guarantees that all efforts to find new life in the 'primordial soup' will be unsuccessful. Without prior injection of information, *nothing* can happen.

Cytoplasmic Machinery

Another unsolved problem is the development of the nano-machines, such as 'chaperones', which fold the proteins, as proteins only become active when precisely folded. The British philosopher of science, Sir Karl Popper, summarised the situation when he wrote: *"What makes the origin of life and of the genetic code a disturbing riddle is this: the genetic code is without any biological function unless it is translated; that is, unless it leads to the synthesis of the proteins whose structure is laid down by the code. But...the machinery by which the cell...translates the code consists of at least 50 macromolecular components which are themselves coded in the DNA. Plus the code cannot be translated except by using certain products of its translation. This constitutes a baffling circle; a really vicious circle, it seems, for any attempt to form a model or theory of the Genesis of the genetic code."*[37] Popper's conundrum is the classic "chicken and egg paradox" and is illustrated in Figure 4.1.

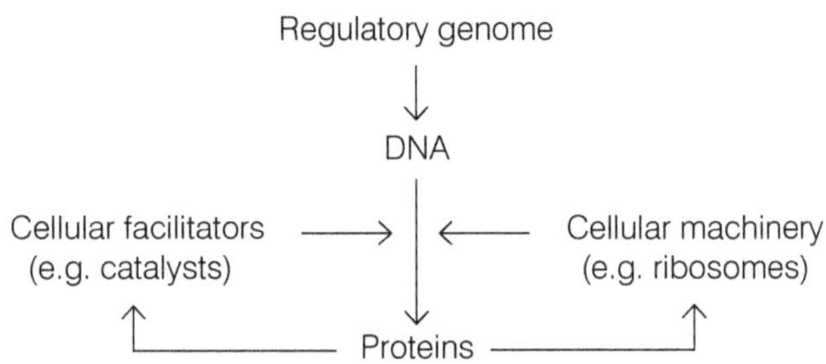

Figure 4.1: Popper's Conundrum: This information flow diagram illustrates Popper's riddle. Popper did not know about the regulatory genome but it only adds to the dilemma. Biologists John Maynard and Eors Szathmary write: *"The existing translational machinery is at the same time so complex, so universal and so essential, that it is hard to see how it could have come into existence, or how life could have existed without it."*[38]

Reproduction (Mitosis)

Although bacterial cell division by mitosis is the simpler form of replication, it still requires more than 50 components and a myriad of intricate steps which must be followed precisely to achieve successful reproduction. Human DNA is nearly two metres long but only twenty nanometres wide so to package it into the cell without tangling requires careful folding. To do this, the cell has spooling machines that wind and unwind the strands of DNA at 8000 rpm.[39] The cell also needs regulatory genetic coding to manage the replication process. This requires a phenomenal amount of specified information yet without it the first cell could neither function nor reproduce.

Law and Chance

Philip Johnson summarises evolutionary theory's problem with information when he writes: *"Genuinely creative evolution... requires a mechanism capable of creating immense amounts of complex specified aperiodic genetic information. Random mutation is not such a mechanism, nor is natural selection, nor is any physical or chemical law. Laws produce simple repetitive order, and chance produces meaningless disorder. When combined, law and chance work against each other to prevent the emergence of a meaningful sequence."*[40] So, although the combination of law and chance is supposed to increase information, it does not introduce new meaningful information into DNA.[41] It adds random variations but even billions of years cannot convert the random variations into finely honed nano-machines if no additional information is added to organise them, so Darwin's theory cannot explain the life we see around us.

Information Management

Human DNA contains approximately 25,000 genes within its 3 billion bases which is at least 10 times more DNA than necessary. The apparently surplus DNA was assumed to be the equivalent of genetic 'fossils' and has been referred to as *'junk DNA'*. However, recent research now suggests that 93% of the genome is transcribed.[42] John Matic, Professor of Molecular Biology writes: *"The majority of the human genome* [is operated] *by a hidden RNA regulatory system that directs differentiation and development."*[43] Thus, information is required to specify the instructions for the correct assembly of the cell too, which is called the regulatory genome.

If we were considering houses instead of organisms, the DNA would be specifying how to manufacture the bricks, floors, walls and roofing materials. The regulatory genome is like the builder who specifies the way the building materials will be connected.

So, in the same way that one builder using identical materials may construct many different houses, a slightly different regulatory genome can build a markedly different body plan using the same materials, the same laws, identical tools and similar construction methods. This explains why there may be up to 98% similarity between the genes in human beings and chimpanzees and yet the body plans are so significantly different. It also explains why the human genome with 25,000 genes has only 28% more genes than a sea anemone.[44]

It is interesting to note, that in human beings only 25,000 genes produce around 100,000 proteins, so there is considerable overlap of the genetic material to produce a fourfold output from the genetic code. This extraordinary feat is achieved by overlapping the gene sequences which can be read in one direction on the first strand of DNA but *in the reverse direction on the mirror image strand opposite!* This is the equivalent to taking a cake recipe, looking at it in a mirror and being able to read the perfect instructions to make lasagne; then cutting and pasting the wording a couple of times to produce a chicken curry, followed by a soufflé! Human beings have not achieved this, even with the aid of powerful computers. So, the idea that such sophisticated gene architecture can occur by random chance, even if it is aided and abetted by Natural Selection, is completely irrational.

The most significant implication of regulatory genes though, is that, because they are almost identical from snails through to humans, the principle of common descent in evolutionary theory predicts that there must have been a shared common ancestor prior to the appearance of snails, which has been dubbed: 'Urbilateria' (the universal common ancestor of all organisms with twofold symmetry). This hypothesised organism must have contained all the control genes for all the higher organisms which were yet to evolve. Professor Sean Carroll wrote: *"The genetic potential was in place for at least 50 million years, and probably a fair bit longer, before large, complex forms emerged."*[45] These genes include the programming for the development of eyes, heart and limbs which Urbilateria did not possess and whose descendants would not use for millions of years to come. Carroll writes: *"It is intriguing to ponder just what so many genes were doing in*

Urbilateria."[46] The answer is that no mechanism exists in Neo-Darwinian Theory for such sophisticated pre-emptory information programming. Darwinian evolution cannot predict the future.

Self-organisation

So, in summary, the formation of the first functioning cell required a staggering amount of information: tens of thousands of genes, proteins and metabolites interacting extensively with one another in order to create the dynamics of a living cell.[47] As randomness and chance are incapable of creating this bewildering degree of complexity, for even the simplest form of life, a number of scientists have proposed that some process intrinsic to the Universe has caused life to appear, inciting the 'laws of nature'.

Two such scientists are chemist and Nobel laureate, Ilya Prigogine and Professor of Philosophy, Isabelle Stengers, who suggested that life might be self-organising.[48] The beauty and symmetry seen in crystal formation and snowflakes are some of the best-known examples. However, although crystals can be extremely large, they are repeating patterns that do not contain a lot of information. In contrast, life requires an immense amount of information to specify it and nearly all the components of life are highly irregular and *asymmetric*. The problem is that self-organising systems do not create novel new information. Stephen Meyer observes that if there were laws or rules for generating information, there would be little variation in information at all. A universal law to produce information would cause regular patterns, such as crystal lattices, rather than result in the generation of the exquisitely varied molecules required for life. Therefore, there *cannot* be laws to cause self-organisation as they are a contradiction in terms. He summarises this point by saying: *"Self-organisational theorists explain well what does not need to be explained. What needs explaining is not the origin of order...but the origin of information."*[49]

Although cells are self-organising, self-regulating, automated machines which auto-construct as well as auto-destruct if the processes go wrong, all these functions must be specified. In addition, complex organisms require embryonic instructions, defining how the organism is to develop from a single cell.

Embryogenesis is the most exquisitely beautiful and complex progression, with a symphony of hormones and molecules orchestrating the development of all the parts of a new human being. This process is so intricate, even the processes have not been fully described to date, let alone an evolutionary pathway for its development.

The Minimum Information required for the First Cell

So, it is very clear that the amount of intricately integrated information in an ordinary cell is immense and I am sure Darwin would have been amazed at the contents of his 'protoplasm'. The first cell had to be completely competent even in its first generation for there to be a second generation. The alternative is the extinction of the species before it could get going. The minimal requirements for the survival of the first cell are shown in Figure 4.2.

From Figure 4.2 it can be seen, that the piecemeal 'learning' dictated by random variation with natural selection, is completely inadequate as an explanation for the integrated information required to form the first cell, because all the functions listed, including the software (black arrows) and hardware (white arrows), needed to be present when that first cell 'sparked' into life. When constructing the first cell for the first time, the bit by bit approach does not work. This falsifies the Neo-Darwinian dictum of gradualism for the origin of life: *'one minute step at a time'*. Life is an all or nothing entity, as it is impossible to be 'partly' alive. You are either alive or you are not.[50]

Information Generation

So, if cells contain so much information, where does it all come from? Since random mutation with natural selection requires an existing cell on which to act, Darwinian evolution cannot explain the formation of the first life. Thus, information was required prior to the generation of the first cell, but that still does not identify where this information came from. Since information is a discrete entity in its own right and needs to be generated by some mechanism, it would be easier if the source had a name, even though its identity is not yet known. The name I have given it is the *'Biological Information Generator'* but science has provided few

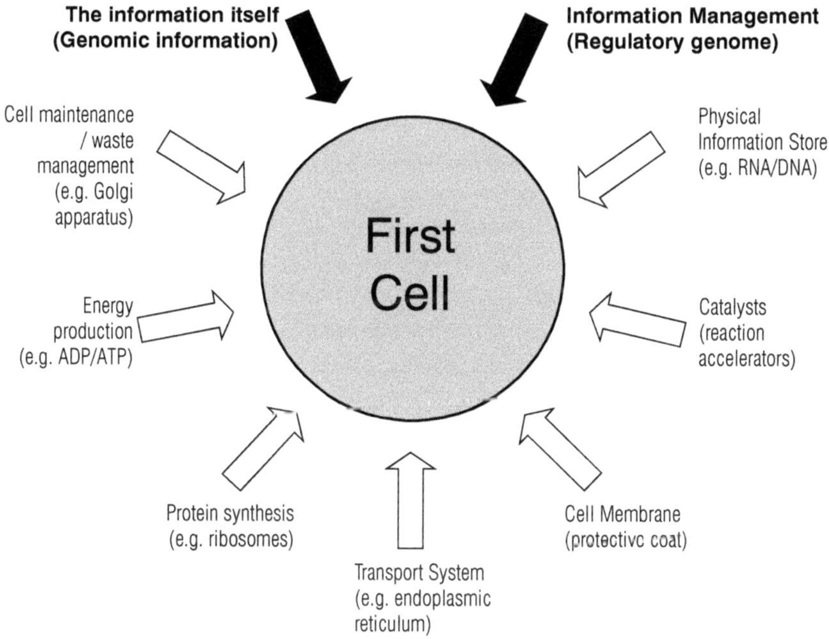

Figure 4.2: The Minimum Information required for the survival of the First Cell Material components of the cell have white arrows and non-material components are in black.

clues to explain the information generation process to date. This will be discussed further in Chapter 7.

The Miracle of Life

In the 19th Century, life was imagined as many simple blobs of protoplasm multiplied many millions of times to create living organisms[51] but with our current 21st Century knowledge of biochemistry and our understanding of the remarkable nano-machinery, there needs to be another explanation for the emergence of the first cell. Microevolution can explain the variation in Galapagos finch beaks but not how birds appeared in the first place (which is macroevolutionary). Similarly, microevolution can explain the resistance of bacteria to antibiotics but not the fact that they exist at all and certainly it cannot demonstrate that finches evolve from simpler forms, such as microbes.

From the above discussion, I would suggest that the original notion of a *'simple cell'* was a misnomer. The *'simple cell'* does not exist and it never could have. Random variation with natural selection may explain the adaptation of an existing species to its environment but it cannot explain the origin of the first cell. Francis Crick commented: *"What is so frustrating for our present purpose is that it seems almost impossible to give any numerical value to the probability of what seems a rather unlikely sequence of events...an honest man armed with all the knowledge available to us now, could only state that in some sense, the origin of life appears at the moment to be almost a miracle, so many are the conditions which would have had to have been satisfied to get it going."*[52] This was written in 1981 but the intervening quarter of a century has only magnified the problem as the previously unimaginable complexity of life has been uncovered.

Professor Chandra Wickramasinghe wrote emphatically in 2000 that the emergence of the first life from a primordial soup on the Earth is merely *"an article of faith that scientists are finding difficult to shed. There is no experimental evidence to support this at the present time."*[53] He goes on to point out that all attempts to create life from non-life, starting with Pasteur, have been unsuccessful. The geneticist and evolutionary biologist John Haldane wrote even more bluntly: *"...we must give up the idea that such an* [initiating] *organism could have been produced in the past, except by a similar pre-existing organism or by an agent, natural or supernatural, at least as intelligent as ourselves, and with a good deal more knowledge."*[54] For the formation of the first cell, meaningful information needed to be introduced in the form of DNA (or a similar molecule) and current evolutionary theory has not even come close to explaining this miracle.

Chapter 5

The matter of Consciousness

Understanding the mind and consciousness (where we interpret meaning) is one of the most challenging subjects facing science today. Bernd-Olaf Küppers' lament is that: *"Phenomena associated with meaning, as expressed in the...dimension of information, appear to evade completely all attempts to explain them on a naturalistic basis, and thus also to escape scientific description."*[1] Indeed, consciousness continues to be such a problem for classical materialism that attempts have even been made to deny that it exists at all.[2] Yet, without the human mind we cannot interpret meaning and our very existence. So, as the materialist approach struggles with these entities, we will explore the information pathway.

Mind versus Matter

Bertrand Russell contemplated the question of mind versus matter writing: *"Is the world divided into mind and matter, and, if so, what is mind, and what is matter? Is mind subject to matter, or is it possessed of independent powers?"*[3] This debate regarding materialism and science has been going on for a long time. Thomas Huxley coined the word 'agnostic' in 1869 as: *"One who holds that the existence of anything beyond material phenomena cannot be known."*[4] But is it valid to assume that reality is exclusively physical?

Adam Frank, Professor of Astronomy writes: *"...after more than a century of profound explorations into the subatomic world, our best theory for how matter behaves still tells us very little about what matter is. Materialists appeal to physics to explain the mind, but in modern physics the particles that make up a brain remain, in many ways, as mysterious as consciousness itself"* Professor Frank points out that some neuroscientists think that they're being precise and grounded by holding tightly to materialist credentials but this conviction is out of step with what physicists know about the material world. This is because at the sub-atomic level the bits of matter are smeared out into sets of potentials or possibilities and all properties of the particle (electric charge, energy, spin, etc) are similarly dispersed. So it is as if the particle does not have definite physical properties at all. The quantum perspective adds a lot of blur to any materialist convictions of what the world is built from and the high ground of materialism deflates when followed to its quantum mechanical roots, because it then demands the acceptance of metaphysical possibilities that seem no more 'unreasonable' than materialism. He concludes: *"...our current understanding of matter alone is unlikely to explain the nature of mind. It seems just as likely that the opposite will be the case."*[5] In a similar vein, particle physicist and Nobel laureate Niels Bohr wrote: *"Everything we call real is made of things that cannot be regarded as real."*[6] Indeed, the physical brain alone is insufficient to explain mind, however, it does appear to be the *platform* for consciousness, so it is the logical place to begin our exploration.

The Brain

Improvements in microscopes and chemical staining techniques during the 19th Century allowed researchers to discover the cells that make up the brain and nervous system (called neurons). Rather than being round like most other cells, nerve cells, were elongated with multiple bumps. These swellings, which formed the connections to neighbouring cells, are called *synapses*. It has since been discovered that all vertebrate brains are composed of neurons and that synapses are the same from lobsters to human beings.[7]

Brain Function and Thought

The human brain is thought to be the most complex entity in the Universe. It is estimated to have 100 billion (10^{11}) neurons, each one of which has been estimated to have between 1000 and 10,000 connections or synapses. The network in the brain therefore has over 10^{14} connections, which all fit very neatly in a space the size of a basketball! Further, for every neurone there are ten supporting cells. In this unimaginably dense and sophisticated web, how do we manage to retrieve anything and when we do, how do we do it so quickly?

Seth Grant, Professor of Molecular Neurobiology wrote in 2009: *"One human brain has a million times more connections then the entire World Wide Web...There are over 200 genes involved with learning and 1500 proteins associated with the normal functioning of a mammalian synapse...This makes the mammalian synapse the most complex of any cellular structure known."*[8] Grant suggested that our brain power is dependent upon this small gap between the nerve cells rather than the cells themselves. These gaps contain the *synapse proteins*. Grant discovered that invertebrates such as lobsters, have about half the number of vertebrate synapse proteins but surprisingly unicellular organisms also contained 25% of mammalian synapse proteins too.[8] This may explain unicellular organisms' ability to process simple information, for example, relating to food sources and light.

In addition, each synapse can utilise between 100 and 150 specialised messenger molecules called neurotransmitters to communicate messages to the next cell; this complexity is far greater than previously imagined.[9] Further still, synapses have the ability to change and adapt in response to stimuli, which is referred to as *plasticity*. This is the process we usually think of as *learning*. To put this into context, when a computer communicates between its microchips, it uses only one type of messenger, an electron, to connect with the next entity in its internal network; imagine the power that might eventuate if your computer's microchips had 100 different electron-type messengers to choose from. This is on top of the 10^{14} connections described by Professor Grant. However, this is only a mechanistic model of brain functioning and gives no hint of how we think. Philip Johnson writes: "[Materialists] *struggle with what is termed the 'hard problem' of mind-science: how it is possible for material events in the brain to produce subjective conscious experience in the mind. If the brain is fundamentally a computer or group of computers, then a sensory input can produce an output (for example, lack of water can stimulate cells to invoke a mechanism to cause drinking). But computers do this without having any conscious awareness."*[10]

So, where do we start to figure out this problem? We know that when we generate a thought, as the neurones converse with each other, they create tiny electro-chemical impulses, the results of which can be graphed as waves on an Electro-EncephaloGraph (EEG) machine. We also know that different regions of the brain are utilised, which can be imaged by Positron Emission Tomography (PET) scanners. Cognitive scientist Professor Marvin Minsky wrote: *"the mind is what the brain does."*[11] The big question is, how does it do it? It is easier to say what mind/memory is *not* rather than what it *is.* Neuropsychologist, Karl Lashley, began experimenting in the 1920s to discover where memory was stored in the brain. He trained rats to run through mazes, then measured how well they remembered as he removed more and more of their brain tissue. Dr Karl Kruszelnicki comments: *"Lashley found that memory is not stored in one single place, but exists throughout the entire cortex, and probably a few other places as well."*[12]

Fast forward to the 1980s, when the British neurologist, Professor John Lorber described treating cases of hydrocephalus, a condition involving the build-up of cerebro-spinal fluid within the skull. In older children and adults this excess fluid makes the brain substance get thinner as the brain is squashed against the bony skull. Remarkably, many young people did not have much brain, but still had normal intelligence. In one extraordinary case, a young man had only a one-millimetre thickness of grey matter in his brain, instead of the average 45 millimetres. Even so, he had an IQ of 126 (the average is 100) and he had gained an honours degree in mathematics![12] In contrast, a stroke can leave us paralysed, so although the entire brain is important for *implementing* actions such as speaking or walking, memory storage and thought can survive partial destruction of many part of the brain and appears to be distributed across it. Our brain 'hardware' is essential for receiving and sending messages to the material world but our thoughts and memory seem to operate differently.

In the same way that information is not the DNA molecule within the cell but an intangible entity that is stored through the ordering of chemical bases within it, EEGs and PET scanning seem to suggest that thoughts are not part of the neurones but are associated with the order or way in which the brain cells are activated. There seems to be some intangible extra that produces thought. Philip Johnson commented: *"If you follow the materialist logic through to the end, ...you will conclude that the personal self in which we put so much stock is no more than an illusion."*[10] Yet we are not an illusion so there must be something more than just the physical brain circuitry and biochemistry.

But what else is there? Could we be looking at the wrong thing or in the wrong place because your thoughts as you read this page are not imaginary? What other evidence do we have? Well, we know that the eye and the ear are frequency decoding devices, which operate digitally,[13] so instead of thought being merely an electro-chemical signal carried by neurones like electrons in an electrical circuit, could it be a wave-like (frequency-modulated) process? Professors Russell and Karen de Valois discovered that the digital inputs from our eyes are converted to frequency information for interpretation by our brain/mind.[14] Thought might therefore be a frequency-mediated holographic process rather than just a physical, mechanistic one. Remember, computers are merely interfaces or translators from our familiar analogue, material world to the digital and back again. Floridi explains: *"a digital interface is a gate through which a user can be present in cyberspace."*[15] Thus, we can live in the material, analogue, space-time domain but using computers we can operate digitally in a non-material, digital domain. Our brains may be functioning similarly, acting as an interface between our material world and a holographic Mindspace domain. However, cyberspace seems to operate via a granular (digital) series of signals which is 'local' in quantum mechanical terms whereas memory follows disseminated, holographic principles, which suggests that brain processing and computer computations are different.

Living in a Multi-dimensional Universe

At this point we need to make a brief detour to examine some other fundamental properties of our universe. Stephen Hawking and Leonard Mlodinow believe that ten-dimensional String Theory and its eleven-dimensional brother, M-theory, will turn out to be the basis for a future grand unifying theory of the Universe. Although this hypothesis is controversial, if Hawking and Mlodinow are right, String Theory leads to some startling implications. If the Universe does exists in eleven dimensions, it would seem logical that there is something in the Universe that exists in at least eleven dimensions and that the formation of the Universe involved these eleven dimensions in some way, even though we can only appreciate and measure the four physical dimensions at present. It would seem possible for human beings to exist in an eleven-dimensional universe, yet not appreciate them all but it is inconceivable that an eleven-dimensional universe could have been created by an entity that can appreciate and exist in less than the eleven dimensions. Therefore any explanation for the formation of the Universe would (as a minimum) involve an eleven dimensional process or entity. However, if an eleven-dimensional universe came into existence, it is reasonable to assume that there is another dimension outside the eleven dimensions, where the universe-forming-entity or process originated prior to the formation of the new universe (Void Zero again). A simple way to illustrate this is to imagine a cartoonist who creates a character, such as Charlie Brown. The cartoonist can write themselves into their comic strip if they choose, but they existed in at least an additional two dimensions outside of the two-dimensional cartoon strip prior to Charlie Brown coming into existence (as the cartoonist lives in, at least, a four-dimensional material universe). So, if the Universe exists in more than the four space-time dimensions, as some cosmologists suggest,[16] then human beings and indeed, all life may also, which could provide a possible solution to the conundrum of consciousness and memory which have steadfastly eluded investigators to date.

Fourier Transforms, Holograms and Memory

We saw above that the brain uses frequency analysis to process visual inputs, which may apply to hearing too. Some scientists have suggested this involves Fourier Transforms to process the information.[14,17] This implies that as our senses take in information from our environment in the physical space-time domain, it is converted into a digital format for processing and interpretation.

> **The Fourier Transform** is a mathematical tool used by engineers to analyse wave patterns. It converts a physical (time-domain) function into its frequency spectrum.

Holograms are also Fourier transforms that distribute a picture (information) across an entire piece of photographic sheet. A standard photograph is a *space-time* image of the object photographed in natural light but a hologram is a *frequency* image of the object that has been photographed using laser light. If the hologram is viewed in sunlight, it looks like a series of blurry smudges and is unintelligible. It can only be seen clearly using the same frequency of light used to create it. If this is done, it produces the strange phenomenon that although the holographic print only takes up the same amount of space as your reflection in a mirror it has three dimensional characteristics such that the viewer is able to peer around objects within the hologram that might otherwise obscure items behind them. Thus, entities within the hologram that were initially hidden can now be seen. These findings are consistent with those of physicist, David Bohm, whose interpretation of quantum mechanics involved a higher dimension which he called the Implicate order.[18] In Bohm's hypothesis, the material Explicate Order is a four-dimensional projection of information that resides in the concealed (folded) Implicate Order in a different domain.[19] The implication is that all the matter in the Universe (the entire space-time domain) is a four dimensional projection of digital information in a (currently) inconspicuous higher domain.[20] Thus, just as holograms are a 2-dimensional (2D) representation of 3D objects, in Bohm's model, 4D objects are a projection of 5D or 6D entities; or maybe up to 11D if M-theory is correct.

Bohm's understanding of quantum mechanics has been less popular than other interpretations over the years but a recent research paper in *Frontiers in Computational Neuroscience* adds weight to this idea by mathematically modelling the activity of neurones that band together to form cliques that create higher dimensional *"cavities"*. Neuroscientist Professor Henry Markram, director of the Blue Brain Project in Lausanne, Switzerland stated: *"there are tens of millions of these* [multi-dimensional] *objects even in a small speck of the brain, up through seven dimensions. In some networks, we even found structures with up to eleven dimensions."* Mathematician Ran Levi of the University of Aberdeen commented: *"It is as if the brain reacts to a stimulus by building then razing a tower of multi-dimensional blocks, starting with rods (1D), then planks (2D), then cubes (3D), and then more complex geometries with 4D, 5D, etc. The progression of activity through the brain resembles a multi-dimensional sandcastle that materializes out of the sand and then disintegrates."* Neuroscience has been struggling to find where the brain stores its memories. Markram speculates: *"They may be 'hiding' in high-dimensional cavities."*[21,22] Could this also provide an explanation for thought in organisms with complex brains? Which brings us to the next topic: life.

What is life?

Edgar Andrews observes that: *"Although proteins and nucleic acids are essential for life, they do not* constitute *life...the creation of a living cell (or a house for that matter) requires ingredients* plus *organisation."*[23] The textbooks say that life is self-sustaining and self-reproducing and it is commonly referred to as the 'Spark of life' but what is it? Richard Dawkins writes: *"The difference between life and non-life is a matter not of substance but of* information. *Living things contain prodigious quantities of information. Most of the information is digitally coded in DNA, and there is also a substantial quantity coded in other ways...* [which are] *the immune system, the nervous system, and culture."*[24] The immune system learns from experiences within an individual's lifetime to prevent further infection (such as chicken pox virus) but it only applies to that individual.

Dawkins' 'third memory' is the one we ordinarily think of when we use the word: the memory that resides in the nervous system that retains a store of past experiences. The fourth 'memory' is the collective teachings handed down from past generations: by word of mouth, in books or on the Internet.[25] But none of these four memories really explains life. All this information is essential for human life as we know it but there is more to life than this. My impression as I studied biology at school was that the genome explained life, however, I have now realised that a string of DNA extracted from a *Tyrannosaurus rex* bone does not make a living dinosaur. The DNA within the bone stores dinosaur information but it is not alive. American biologist, Professor Stuart Kauffman wrote: *"autocatalysis and molecular reproduction are necessary for life, but not yet sufficient. Life possesses deeper, still more mysterious properties."*[26] Life also requires a non-material factor which is more than just the physical components of a cell, because to create life requires *dynamic* information which becomes active within a living cell.

Richard Dawkins makes the observation that: *"DNA doesn't care about being 'dead', it is just coded information."*[27] These are some of the problems associated with defining life, so the question: 'What is life?' may be easier to answer if we could first decide what defines the absence or loss of life, which is death.

What is Death?

Life requires matter, energy and information but we have seen that on their own they do not create life. Matter, energy and information can be used to create a computer which functions and can produce real outputs. It is also possible to visualise the construction of a factory that would allow our computer to make more copies of itself, equivalent to replication but does that confer the status of 'life' on the computer? There is something else, call it a 'life factor' that distinguishes a computer from a living organism. It also distinguishes a living organism from a dead one. When an organism dies, five minutes after life becomes 'extinct' the body and DNA are physically the same in the space-time domain but 'life' is missing.

So, what is death? There are different types. Cell death is when the chemical processes in the cell irretrievably break down. The second type may occur if a person has a heart attack and their heart stops beating, the person loses consciousness within seven seconds due to the loss of an oxygen supply to their brain, as the brain cannot store oxygen. If the heart is not started again within ten minutes, then 'brain death' occurs. However, if the person's heartbeat is returned to normal within the first few minutes, it is possible for the patient to make a full recovery. In severe head injury or near drowning, sometimes the victim's physical body survives but the person is declared 'brain dead'. They have lost their consciousness, their personality and all the essential qualities we think of as 'human' but their physical body still functions perfectly (sometimes for many years if they remain on a ventilator) and their organs (such as their kidneys) are often able to be successfully transplanted into someone else and survive for decades. Yet they are 'not there', so are they alive or dead? Their physical body is 'living' (at least the cells are) but as human beings they are, for all intents and purposes, 'dead' as they cannot function like a 'living' person. There are therefore at least two types of death: cell death and death of the consciousness. Cell death can be seen when the blood supply to a limb is lost, such as in a smoker who develops gangrene of a toe from blockage of their arteries to that foot. Their brain is still completely normal, but some of their peripheral cells have died, their toe turns black and needs to be amputated.

Thus, there is a stream of consciousness that is mass-less and non-material that is distinguishable from our physical body. We know that this consciousness is dependent on the physical network of nerves in the brain, in the same way that a computer program needs its computer circuitry (hardware) to operate but consciousness is more than the hardware. So could these thoughts and consciousness reside somewhere else, perhaps Markram's 'high-dimensional cavities'? A mechanism is not yet known.

When brain death occurs, some vital ingredient is lost from the living body but what is it? One clue is that consciousness (which is just as difficult to characterise as 'life') is lost as well. Much research has gone into trying to determine if life and consciousness are emergent properties of the functioning body and brain, such as the magnetic field that is generated when current flows in a completed electrical circuit; or are life and consciousness extrinsic properties, that are formed before birth and lost at death, such as the software programme in a computer?

Weighing up the preceding information, cell death appears to involve biochemical breakdown that results in the loss of material activity, whereas the death of consciousness appears to involve the loss of a non-material entity, as a human being who is 'brain dead' on a ventilator is physically identical, yet mentally different to their former healthy self. Observe also that the memory of the cardiac arrest victim who is successfully resuscitated carries on from where they left off just before they blacked out. Thus the 'life' that we observe in an amoeba appears to be a dynamic (material) process and has emergent features that are dependent upon the physical integrity of the cell and the activity of the intrinsic information in its DNA, whereas the phenomenon of consciousness that is seen in more complex organisms, such as mammals and human beings requires the input of an additional information system. This information or entity, despite extensive searching for over 90 years, has not been found in our physical realm, so it would appear to be in another place; let's call it the 'fifth dimension' to give it a name for discussion purposes. If that is the case, then it can be inferred that at the death of the organism, information in another dimension might become separated from its physical hardware platform in dimensions one to four. The physical body of the deceased person retains its genetic information in the space-time domain, which is in the first four dimensions, but it has lost the 'spark' of life, along with the molecular activity within the formerly living cells. In teasing apart the above mechanisms of death, it can be seen that cell death is a hardware problem whereas death of the 'consciousness' is an information or software issue.

Thus, as a cell must already be living when a new life materialises at conception, the 'spark of life' is handed down from one generation to the next. If there is an extra dimension housing consciousness, when the gametes are formed from the living, parental cells, they would bring the extra dimension that has the potential for consciousness with them, in the same way that our third dimension is inextricably linked to our first two dimensions. In our three spatial dimensions, you cannot physically make yourself into a two-dimensional being, although you can create a two-dimensional image, such as a photograph; however, that is not alive. When the sperm and ovum unite, they form a new competent life in our space-time domain, with the fifth dimension that has the prospect of consciousness attached. Before the union of the sex cells, each gamete has half the potential for new life, but being incomplete, it lacks its operating platform (the physical brain) so cannot yet be conscious.

If a person dies, their physical body and brain in our material domain loses its 'spark of life' while the entity that is in the fifth dimension, which is consciousness, disappears from our view, as we can no longer access it from our space-time domain. But since this information defining consciousness is massless and hence timeless, it may persist unchanged in its non-material location, as pure information. Separated from our physical world, consciousness presumably does not age. Hence brain and mind are very different.

Consciousness

Consciousness, or awareness of our surroundings, is such an elusive concept that most of us rarely contemplate it but we use it all our waking hours and it needs an explanation. Yet science is currently at a loss to even know where to start. Oliver Sacks, Professor of Neurology, lamented that despite all our advances in neuroscience, *"'...neurological activity and psychic activity still seem utterly different,' even 'incommensurable' in kind."*[28] Andrew Gosler of Oxford offered some insight when he wrote: *"Information exists most powerfully in the mental realm; it is surely the very stuff of consciousness..."*[29] Consciousness is like an operating system on a computer; it is a platform on which other useful programs, such as thoughts and feelings run. Without this software package there is no useful output or productivity from the computer. Conversely, the hardware system by itself yields nothing either; it is the integrated hardware and software package combined into one active system that is productive – once again, layer upon layer of information!

Further, Christof Koch, Professor of Computation and Neural Systems, points out that most of what we do is not under our conscious control. Nor are we *aware* that we are doing it. *"'When you speak, you do not have to think about grammar and syntax and vocabulary, you just open your mouth and the words come out. 'If you had to consciously think through that stuff, ...you'd never get anywhere.'"*[28] Speech relies on the competency of both the conscious and the sub-conscious parts of our brain to function harmoniously. In the same way, to type this paragraph on my computer I needed both a fully competent operating system and a word-processing package, yet neither of these software tools was responsible for the words that you are reading. Remember that the electro-chemical messages in the brain and the software packages in the computer are simply vehicles that enable processing of the information.

However, in principle, Koch says, there is no reason why consciousness is necessary for living creatures. With enough sensors and programming it is conceivable that a robot could do almost anything. The problem is that every robotic behaviour must be hardwired and the more situations it must respond to, the more complex its programming becomes. *"Instead, ...a much more powerful and flexible system [is]that [which] we call consciousness."* Consciousness enables organisms to deal rapidly with unexpected events and to plan for the future.[28] Without consciousness we have to rely on pre-programmed information as computers, insects or robots do, which removes any form of spontaneity or free will,[30] which we treasure as a hallmark of being human. Koch's analogy with robots highlights the significance of consciousness. Table 5.1 on the next page compares these two systems.

So, consciousness is a fundamental quality that makes us human but where does it come from? There does not appear to be a 'consciousness centre' within human beings, but consciousness appears to be the result of global brain activity, which suggests that consciousness is 'distributed' throughout the brain.[31] Thus, as previously seen with memory, consciousness also exhibits holographic properties.

All mammals and birds appear to exhibit consciousness, but we will begin by considering the *human* brain, as there is no doubt about the relationship. The mental abilities of other organisms will be considered in Chapter 6. The human brain with its 100 billion

	Computers	Human Brain
Hardware	Computer components	Nerve cells and supporting tissues
Software	Machine Language Operating system Applications packages User input (extrinsic)	Biochemistry Consciousness Speech Thoughts (intrinsic)
Mechanics	Analogue to digital interface (cyberspace)	Analogue to holographic interface (mind)
Decision making	Pre-programmed (deterministic)	Free will / self-will (choice)

Table 5.1: Information processing in Computers versus the Human Brain
Note that all software, whether for computers or living organisms, is ultimately information.

neurones stands out in its ability to look beyond mere survival. Chimps can plan over a short period of time, such as hours or days, but only humans can plan for years, decades or centuries into the future. The human brain is three times the size of a chimp's brain, although brain size alone does not determine intelligence. Women's brains are almost universally smaller than men's but there is no verified data demonstrating a difference in the level of intelligence between men and women. Instead intelligence appears to depend upon the total number of connections between the individual nerve cells.

So, is consciousness purely an emergent property of electro-chemical brain activity running on biological hardware or is there more to it? Alan Gijsbers, Professor of Medicine wrote: *"George Engel's biopsychosocial model of understanding humans, challenges the simple biomechanical approach to patients. Engel describes humans as a series of emergent systems from sub-atomic particles, through cells, organs/organ systems to the person... Engel uses the concept of emergence, whereby, as we move to the next level, new properties emerge which could not be predicted from understanding the state of the lower level. A useful slogan for emergence is, 'We cannot experience wetness by looking at the isolated H_2O molecule' because surface tension comes about by the interaction of many water molecules together."*[32] But how might this emergent property which causes consciousness work? Professor Gijsbers continues: *"Neurological diseases illustrate the tight link between the mind and brain, and we recognise that defects in brain function (Alzheimer's, strokes, tumours etc) have a profound effect on mind function. But mind functions can also affect brain function..."*

"The trouble started with Descartes. ...He tried to work out what were the most basic axioms, and from there deduce the full system via pure reason. ...Descartes subjected all he knew to radical doubt; what was left in the end? That he existed and that he thought. He therefore developed his famous axiom, I think therefore I am. ...Further, his thinking world was separate from the physical world. The world of the mind is the world of the soul is the immaterial world of the spirit. This was of a completely separate substance than the physical substance of the material world. Thus he entrenched dualism into our thinking. This did allow physicians to study the body as a mere machine, but it also forged the separation between the world of the mind, soul spirit and the world of the body or the environment. We are still grappling with Descartes' legacy - both its benefits and drawbacks."[32]

Hawking and Mlodinow add: *"Descartes, for instance, in order to preserve the idea of free will, asserted that the human mind was something different from the physical world and did not follow its laws. In his view a person consists of two ingredients, body and soul. Bodies are nothing but ordinary machines, but the soul is not subject to scientific law."*[33] Hawking and Mlodinow who profess belief in scientific determinism continue: *"It is hard to imagine how free will can operate if our behaviour is determined by physical law, so it seems that we are no more than biological machines and that free will is just an illusion."* However, the fact that something is *'hard to imagine'* does not make it untrue. Einstein's theories were not conceived of before 1900 but now no physicist doubts them. Hawking and Mlodinow are forced into their conclusion because of their presupposition that physical law is the ultimate entity in the Universe. It is not surprising therefore that they write: *"While conceding that human behaviour is indeed determined by the laws of nature, it also seems reasonable to conclude that the outcome is determined in such a complicated way and with so many variables as to make it impossible in practice to predict."* So, having stated that they believe that free will is an illusion, on the next page they write: *"In the case of people, since we cannot solve equations that determine our behaviour, we use the effective theory that people have free will. The study of our will, and of the behaviour that arises from it, is the science of psychology."*[33] By assuming that scientific determinism is the only possible answer, Hawking and Mlodinow have forced this contradiction upon themselves.

> **An effective theory** in physics is a framework created to model certain observed phenomena without describing in detail all the underlying processes.

Gijsbers expresses an alternate view: *"If on other grounds we believe in an unseen world acting (somehow) on the seen world, then our failure to see how that should be, does not rule out that the unseen world can act on the seen world. If we do not understand how the mind affects the brain, it does not mean the mind does not affect the brain."* Neuroscience has not removed the need to consider higher-level phenomena, even if we do not fully understand the relations between the layers. Gijsbers concludes: *"Mind activity, while dependent on brain activity, cannot be reduced to brain activity. Even reductive scientists would agree to such a statement."*[32]

So, since the purely reductive physical model cannot explain our observations, is Gijsbers' suggestion of parallel *"seen"* and *"unseen"* worlds any more productive? There is no generally accepted model for consciousness, although scientists, such as Stanford neurophysiologist, Carl Pribram have suggested a holographic model of mind while Michael Talbot proposed the integration of Pribram's ideas with David Bohm's interpretation of quantum mechanics. For if the physical universe is but a secondary reality and what is 'there' is actually a holographic blur of frequencies, although we may think we are physical beings moving through a physical world, this may be illusory, like a digital simulation in a computer game.[34] A third possibility is that we might have both a physical body with a holographic mind. This description may initially seem incongruous because we are not used to thinking of ourselves as existing in more than four dimensions. We are familiar with our material state which we consider is the 'real' world and which Bohm termed the Explicate Order but as Gijsbers points out, it does not rule out an information mediated domain, which Bohm termed the 'Implicate Order'. Each mirrors the other, like the image of a landscape reflected in a lake. If our existence in the Universe is as a multi-dimensional entity, as String Theory proposes, then we have both a material, physical presence and an illusory (but still real) digital and/or holographic identity in another domain but they are not completely separate. In the same way that our depth cannot be separated from our height or width, while we are still living, our digital dimension is joined intimately to our familiar physical dimensions. The separation only occurs after death with the loss of the physical body.

A Compound Model of Mind

In the same way that an appreciation of death assisted in understanding life, so the study of unconsciousness can be helpful in gaining a better understanding of consciousness. In medicine we not infrequently see patients who have experienced a period of unconsciousness after head injuries. Although the mechanism is not understood, it is presumed that consciousness is lost due to temporary damage to a significant number of neurones, which reduces the active network to less than the threshold required for consciousness. When the unconsciousness is prolonged it is called coma. In coma, the person is still alive but not conscious, as is also seen in sleep and anaesthesia. Thus, the state of consciousness is independent of the state of being alive. As discussed above, Engel sees consciousness as an emergent property of our physical state but a detailed credible mechanism is lacking and Engel's model alone seems incomplete. Alternatively, if consciousness is viewed as an interaction between an information-specified, non-material mind (the software component), that is independent of the physical space-time domain, yet requires a sufficiently large brain cell network on which to operate (the hardware component), then a credible mechanism begins to emerge, which we will call the Compound Model of Mind. Free will can then be understood as the product of a conscious, non-material mind which generates choice while the mechanistic principles of the material domain result in determinism for non-conscious entities such as insects. This Mindspace domain is illustrated as an extension to the Information Map of the tangible universe in Figure 3.2.

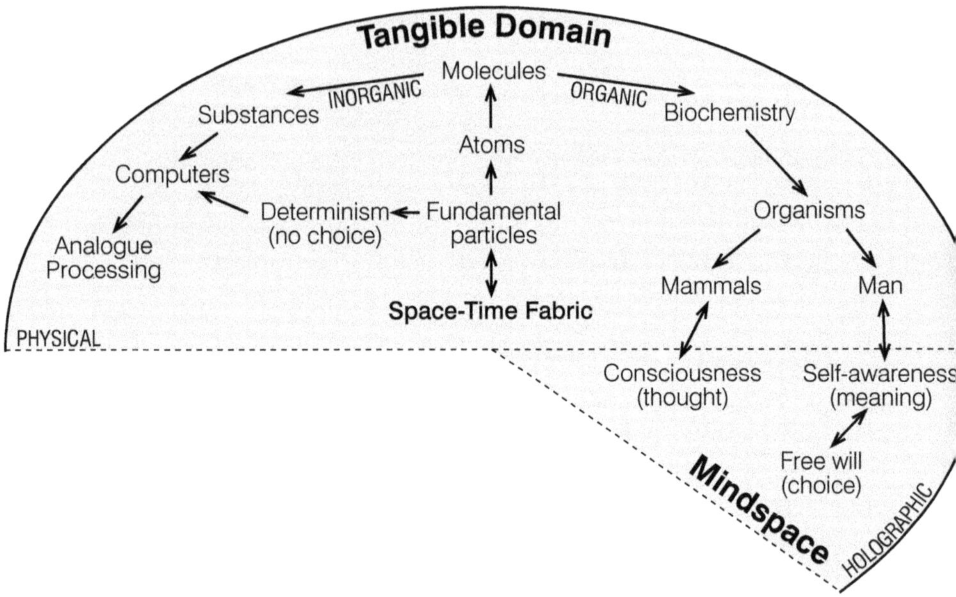

Figure 5.1: Mindspace Information Map
Mindspace has holographic, non-material properties, so consciousness and self-awareness appear to exist in a different domain to the physical brain.

Chapter 6

Mind and Meaning

The Mind

As already seen, no physical basis has been found to explain the human mind. Nobel laureate in Medicine, Christian de Duve wrote: *"...what appears to me as supremely significant is that a combination capable of giving rise to life and mind should exist at all."*[1] We previously discussed consciousness as having similarities to a computer's operating system but now we need to address the higher function of self-awareness through which we appreciate meaning.

Edgar Andrews notes that as far as we know, only man possesses 'mind' – the capacity for self-knowledge and abstract thought; nothing about man is more unique than these mental powers – abilities that represent an enormous gulf between Homo sapiens and every other species.[2] But describing the differences is the easy part, explaining how it works is the difficult bit. The choice is between monism and dualism. Monism denies that there is any real distinction between mind and matter and claims that mind is simply the brain at work. In contrast, a dualist model, where the non-material mind is discrete from the physical brain was developed by Descartes who, as we discovered in the previous chapter, began the modern debate that continues to grow into a vast literature.

The atheist philosopher Bertrand Russell presented the consequences of the monist view in 1903 when he wrote: "...

even more purposeless, more void of meaning, is the world which science presents for our belief...That man is the product of causes which had no prevision of the end they were achieving; that his origin, his growth, his hopes and fears, his loves and his beliefs, are but the outcome of accidental collections of atoms;...all these things, if not quite beyond dispute, are yet so nearly certain, that no philosophy which rejects them can hope to stand. Only...on the firm foundation of unyielding despair, can the soul's habitation henceforth be safely built."[3] The problem with the monist view is that mind becomes meaningless. If our thoughts are purely the product of mechanistic processes dictated by our genes, then our reasoning is not our own and we cannot know if our thoughts are valid or not. The avowed rationalist, J.B.S. Haldane confessed the monist's dilemma when he wrote: *"If my mental processes are determined wholly by the motions of atoms in my brain, I have no reason to suppose that my beliefs are true. They may be sound chemically, but that does not make them sound logically. And hence I have no reason for supposing my brain to be composed of atoms."* The problem then arises that if our brains are genetically preconditioned to produce only specified forms of thought and behaviour, the antisocial person is no longer responsible for their actions but is merely the victim of their genetic inheritance. It is simply their genetic predestination and it would be morally wrong to punish them for it.[4]

So which theory is true? What evidence do we have? The surprising discovery of the last decade is that when we learn a new task, we grow new brain cells from resident stem cells and/or re-wire the existing neurones.[5] This process takes six weeks. Thus, *active thought that results in learning*, changes our physical brain structure. So, it is not the material structure of our brains that determines our thoughts but completely the opposite. Our genetic inheritance does determine the physical components of our brain hardware, creating consciousness but after that our mind is in control. Ultimately it is our thought processes that determine the way our hardware is networked. These findings vindicate dualism

> **Stem cells** are pluripotent meaning they can change into any type of specialist cell as needed.

while contradicting monism, because if thought was purely a by-product of our physical brains, there would be no way that thoughts could retro-wire the physical brain structure.

Self-awareness

No one doubts that self-awareness is a trait possessed by all normal human beings.[6] However, are chimpanzees, which are generally considered to be the next most intelligent species after man, self-aware? The evidence would suggest that they are not. They do not wear clothes or apparel for modesty (whereas almost every human people group does). Neither do they appear to reflect on their own existence or death; no chimps have been observed to practice funereal rituals or create memorials to the dead as humans do. Although primates can be observed to be thinking and able to perform simple problem-solving exercises, there is no evidence that they possess self-awareness. Without an ability to self-reflect, animals are not considered morally responsible for their actions; human beings can be tried for murder but sharks, crocodiles and grizzly bears cannot.

Robin Dunbar, Professor of Evolutionary Anthropology, Oxford observes that despite all the fuss being made about cultural traditions in chimpanzees, orangutans and whales, only one species has built the Great Wall of China, put men on the Moon, thought up Pythagoras's theorem, conceived the existence of life after death, written the plays of Shakespeare and composed the music of Johan Sebastian Bach. *"In short, it is in human's capacity for culture, our capacity to live in a world constructed by ideas, that we really differ from the other apes…"* Dunbar cites the phenomenon known as 'theory of mind' – the capacity to imagine another individuals 'mind states' – as a critical difference between humans and other animals. *"It is widely accepted now that most higher animals have knowledge of their own mind states. In terms of philosophy of mind, this is what is usually referred to as first-order intentionality. 'Intentionality' is a general term covering those mind states that involve beliefs. They are evidenced by the use of words like believe [that something is the case],* know, suppose, intend, want, desire, wish, *etc. They imply that we have a certain level of awareness about the contents of our minds, that we are aware that we believe...that something is the case."*[7]

First-order intentionality is the world inhabited by very young children. It is not self-consciousness but is certainly a form of consciousness that is shared with birds and mammals. However, around four years of age, children develop the capacity to understand the minds of others as well. This capacity is referred to as 'theory of mind'. Once they have acquired the capacity for theory of mind, they are able to engage in imaginative play, that a doll can drink tea from a teacup, even though the cup is empty, and they can lie with conviction.

> **Theory of mind** is literally intended to mean that the person has a theory about minds, a theory about other people's capacities to have beliefs.

"With theory of mind in the mental toolkit of young children, they become good psychologists – they understand the mind behind the behaviour and begin to understand how false information can mislead that mind. They are in what philosophers of mind would call second-order intentionality – they believe that someone else believes [that something is the case]. Before this, they have always assumed that everyone else shares the same knowledge of the world as they do." Dunbar concludes: *"There is no strong evidence to suggest that any species of animal other than humans has theory of mind."* Yet in human beings, babies have first-order intentionality, five-year-olds have second-order and normal adults scale in at fifth-order.[7] (See Table 6.1.)

The fact that birds, mammals and human beings have consciousness but that only the latter possess self-awareness, suggests that there is a category difference, rather than being the different ends of a spectrum, as there is a chasm of gigantic proportions between the cognitive abilities of chimpanzees and adult human beings. If they are indeed discrete and consciousness resides in a fifth dimension as discussed in Chapter 5, then I would suggest that self-awareness resides in a 'sixth' dimension.

Understanding thought within a multidimensional universe is challenging, and most of us have never thought about this before,

Domain	Dimensions	Active Mind States	Examples
	1-4	Life without consciousness.	Plants; simple life (e.g. bacteria); sleeping complex animals.
	1-5	Life without self-awareness: simple thought and emotions.	Complex animals (birds and mammals); human infants.
	1–6	Life with self-awareness.	Awake adult human beings.

Table 6.1: Characteristics of the Domains of Life
A study published in 2009 suggested that laughter is a universal language which is shared by all humans, primates and probably mammals. This suggests that mammals experience feelings and emotions too and so emotion would reside in the fifth dimension.[8,9]

but an analogy may bring these ideas closer to home. If you access your bank account online using an electronic banking facility, the information is stored on one or more host computers somewhere in the World but the information, as far as you are concerned, is 'out there in the cloud'. You only need a suitable interface, such as your smart phone or computer. In the same way, could it be that if we are multi-dimensional beings, that we are able to work simultaneously in different domains, some in the physical space-time domain and others in a non-material domain? The problem is that we do not know how the interface might work yet.

So, what happens to our self-awareness in the sixth dimension when we die? If the electrical circuits in a computer burnout and the hardware 'dies', the software malfunctions also. Your bank account still exists in another place (in computers across the internet) but it is inaccessible by this route. A connection to the bank's software can be restored if your computer hardware can be repaired (equivalent to cardiac resuscitation), but, if not, a new computer must be purchased (equivalent to a new physical body in the first four dimensions), with functioning machine language software (life) and a new connection must then be established to the bank's software over the internet (consciousness in the fifth dimension). In living things, however, as only life begets life, there is no current way to reconnect to a lost consciousness from a body in the first four dimensions. It is conceivable that in the future it will be possible to intentionally do so but that will require a better understanding of the interface between the two. With our present knowledge of science this sounds farfetched as these areas are poorly understood but progress might be made by examining other entities that share this interface, such as sleep, dreaming, thoughts and memory.

Sleep and Dreaming

Sleep is a temporary absence of consciousness, so if consciousness operates from a fifth dimension, the interface controls would seem to vary when sleep cycles occur. We know that some natural hormones, such as rising melatonin levels in the evening, precede sleep and rising cortisol levels early in the morning precede waking, so there are measurable physical changes associated with sleeping and waking but no mechanism has been described yet. Physical tiredness and thought are usually involved too, so once again, hardware and software are woven together for normal functioning.

Studies of brainwave patterns show that there are three different types of sleep: deep sleep, light sleep and dreaming sleep. A typical, healthy, adult human being has an average of three hours deep sleep, three hours light sleep and two hours dreaming sleep each night but our periods of slumber are divided into alternating segments of the various types. Each phase of sleep lasts between twenty minutes and one and a half hours. Deep sleep is when the body appears to partially shut down, to physically rejuvenate its cells in the space-time domain. Dreaming sleep, which is also called rapid eye movement (REM) sleep, allows us to replay past events where we can revue, explore possibilities and problem-solve; this rejuvenates the mind in its non-material domain. Light sleep appears to transition between the other two types of sleep and acts like the neutral gear in a car's gear box, allowing smooth transitions from wakefulness to sleep and then in and out of deep and REM sleep; waking directly from REM sleep risks sleep paralysis. As mentioned previously, electro-encephalograms (EEGs) chart our brainwave activity in the same way that an electrician can measure electrical voltage potentials across your computer's circuitry but just as the electrician cannot use a voltmeter to identify the contents of the electronic documents on which you are working with your computer, the EEG cannot identify what you are thinking. That would require a 'thought analyser' but if thoughts reside in a higher dimension, that would be beyond the measuring capability of an EEG machine recording events in our physical domain.

There are many intriguing questions about sleep. An obvious one is: what causes the switching between the different types of sleep? It is interesting that in dreaming sleep, the mind appears to be disconnected from the physical body and one can have the sensation of trying to run away in a dream but our legs will not physically move until waking, when we suddenly find ourselves thrashing in the bed, as the physical brain and mental processes are re-engaged. While dreaming, our eyes are active as they are direct extensions of the brain. Deep sleep appears more akin to coma, where there is minimal mental activity; in this state a person is much harder to rouse as the physical brain is in temporary shut-down mode.

Thoughts, Feelings and Emotions

Thought may be defined as a process that occurs in a living brain, with the ability to generate or manage information. It is an information manufacturing, manipulation and distribution mechanism but it is very poorly understood at present. As a result, the potential of thought is probably underestimated. The BSI-Toyota Collaboration Centre in Japan has developed a brain-machine Interface that allows the elderly or disabled to use signals from their brains to control a wheelchair in as little as 125 milliseconds.[10] Thus our non-material thoughts are able to physically change our immediate environment. The big question is: where and how do thoughts happen? A thought generation centre has not been identified in the material brain so maybe they are initiated in the non-material sphere while being executed through the physical brain. If that is so, then our thoughts are still within our brains but happening in a non-material way. (See Table 6.2 on the following page.)

To date, no specific location within our brains has been identified for the physical *storage* of thoughts and feelings either, however we do know that certain functions are *processed* in

Component	Function	Example
Physical (hard-wired) brain	Information input/output & matching	Insects
Conscious brain	Information recognition & first-order analysis	Dogs
Self-aware brain	Complex analysis, generation of novel information and ascribing meaning	Human beings

Table 6.2: Processing levels within the Brain

different parts of the brain. Sleep is initiated in a localised part of the base of the brain while vision is interpreted at the back of the brain and speech initiation is processed in specific parts of the left hemisphere. Lessons have been learned from localised brain damage, such as that sustained from a shrapnel injury to a Vietnam veteran's right frontal lobe, which left him lacking insight and unable to plan. These types of injuries have allowed scientists to map the various functions of different regions of the brain.[11] From this type of research, it is clear that wherever thought resides, a fully competent physical brain architecture is necessary for normal brain functioning but that the physical brain alone is insufficient to adequately explain thoughts. At this stage, analysis of the mental abilities of other living organisms can assist our understanding, as summarised in Table 6.3.

In Table 6.3 it can be seen that simpler organisms such as viruses, bacteria and plants do not display features suggestive of higher dimensionality. Insects are difficult to place, as their capacity for sleep is still uncertain, whereas crustaceans, fish, amphibians and reptiles all display some characteristics of thought. Birds, mammals and primates demonstrate additional characteristics of cognition but it is unclear if this is hard wired (instinct) or soft wired (reasoning) or both. Human beings alone demonstrate self-awareness. Professor of Biological Anthropology and Neuroscience at Berkeley, Terrence Deacon notes that there are: *"...hundreds of thousands of species with brains, and tens of thousands with complex behavioural, perceptual, and learning abilities. Only one of these has ever wondered about its place in the world,... "*[15] Thus self-awareness in human beings operates at a different level to thoughts and memory and it is therefore

elds	1			2					3
mension	1–3rd	4th		5th					6th
aracteristic / nus	Space	Time	Life	Thought & memory	Deep sleep	Hibernation	REM sleep/ dreams	Reasoning	Self-awareness
cks	√	√	x	x	x	x	x	x	x
uses	√	√	≈	x	x	x	x	x	x
cteria	√	√	√	x	x	x	x	x	x
nts	√	√	√	x	x	x	x	x	x
ects	√	√	√	≈	≈	x	x	x	x
ustaceans	√	√	√	√	≈	x	x	x	x
h	√	√	√	√	√	x	x	x	x
nphibians	√	√	√	√	√	√	x	x	x
ptiles	√	√	√	√	√	√	x	x	x
ds	√	√	√	√	√	x	√	√	x
mmals	√	√	√	√	√	√ (+/-)	√	√	x
mates	√	√	√	√	√	x	√	√	x
mans	√	√	√	√	√	x	√	√	√

Table 6.3: Mental attributes according to various categories
Although higher dimensions are not directly measurable, we can gain some understanding by comparing different organisms. Insects and crustaceans demonstrate some features of the deep sleep seen in mammals but they differ in other respects.[12,13,14]

not unreasonable to conclude that it is in a unique class when compared with similar characteristics in all other species.

These differences extend to artificial intelligence too. Chemist and philosopher, Dr J.P. Moreland observes: *"...computers have artificial intelligence, not intelligence. And there's a huge difference...A computer has...no awareness, no first-person point of view, no insights into problems ...Augustine said animals have thoughts, but they don't think about their thinking. And while we have beliefs about our beliefs, animals don't."*[16] And neither do computers! Computers are said to operate in cyberspace which maybe the environment of the microchip hardware or somewhere else but it appears to involve a non-material process that is 'local' and hence different to Mindspace discussed in the previous chapter. The Cyberspace domain is mapped in Figure 6.1.

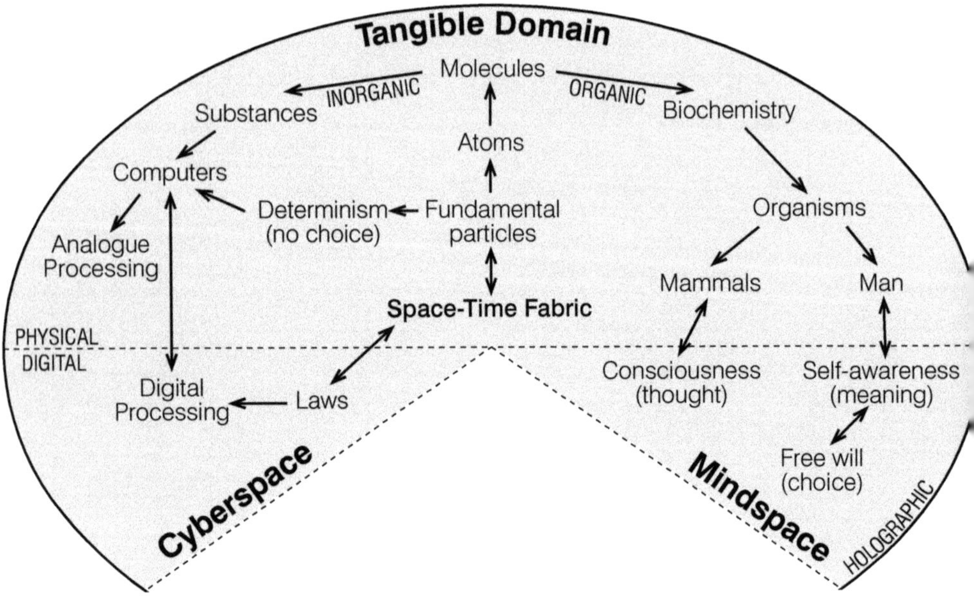

Figure 6.1: Information Map of Cyberspace
Cyberspace performs deterministic processing, as opposed to mind which has freewill.

In contrast to Cyberspace, Mindspace provides self-awareness and choice (freewill), which allow the phenomenon of conscience. Conscience involves the concept of being others-centred, such as the ability to know the difference between 'right' and 'wrong', which is a uniquely human trait. It is not the knowledge that there will be negative consequences for certain actions (dogs know this), it is doing something that is in the best interests of someone else, *for no personal gain.* Helping an elderly person across a busy street would be a typical example. This is called *altruism*. The question is: from where does this arise? Even Richard Dawkins admits that Darwinian Theory has no answer. Physicist, David Bohm believed that thoughts and behaviour do not derive from our physical domain alone but that: *"Every action starts from an intention in the implicate order."*[17] This is logical if our thoughts operate from a non-material domain. But now it is time to look at the cumulative effects of thought, which is memory.

Memory

Memory is the ability to record, store and recall thoughts. In contrast to thoughts, which appear to be dynamic information, memory seems to be an information warehouse. Memories can generally be recalled at will, which is an amazing ability when we consider the vast amount of information encountered in a single day, let alone a lifetime. Physicist and mathematician, John von Neumann, calculated that over the course of the average human lifetime, the brain stores 2.8×10^{20} bits of information.[18] We have seen that thoughts are associated with an electro-chemical signal that can be imagined as a 'brainwave' and can be detected by an EEG, so thought is associated with a moving signal in neurones. In contrast, memory is a stored thought and cannot be imaged by any known modality. It appears to partly involve the pathway taken by the signal through the network of nerve cells but what directs the switching of the signal at each nerve cell? We know that the transient electro-chemical signals are converted to some other form of long-term memory with the assistance of the mammillary bodies, which are located on the base of the brain, because people who have severe damage to their mammillary bodies, such as after a head injury, can no longer lay down long-term memory. They can remember old memories prior to the injury and remember new events for up to two minutes but cannot generate new long-term memories.

But memory is only the tip of the iceberg! When we attempt to understand brain function there are many other aspects that need to be considered, including creative thought, personality and logical analysis but we have been hampered to date by what we can objectively measure. The creative output of the brain is orders of magnitude greater than that which our current understanding predicts from the input and processing that we detect within the physical brain. The closest analogy we have is a computer network, such as the internet, yet computer networks do not come close to the original thought, creativity or critical analysis that all normal human beings are capable of performing. More than just the *communication* between individual nerve cells, there is a very powerful production process that generates new original information. Phillip Johnson wrote: *"What makes the mind from the brain is not the neurons but the information that coordinates the neurons and uses them (and perhaps other entities as yet unknown) to produce the phenomena of our mental life."*[19]

Although much is still unknown, there is some research to direct our quest to understand the human mind and memory better. In the 1920s, Canadian neurosurgeon, Wilder Penfield electrically stimulated the brains of epilepsy patients and found he could cause them to move their arms or legs, turn their heads or eyes, talk, or swallow yet the patient would respond by saying: *"I didn't do that. You did."*[20] Penfield concluded that patients consider their mental selves as having an existence separate from their bodies and there was no place where electrical stimulation will cause a patient to believe or decide.[21] Thus, Penfield could control his patient's physical actions with electrical stimulation but not their thoughts; thoughts are beyond the physical.

Meanwhile, in the US, psychologist Karl Lashley found that no matter what portion of a rat's brain he removed, he was unable to eradicate its memory of how to perform complex tasks it had learned prior to surgery.[22] These findings were confirmed in the 1960s by Carl Pribram, a neurophysiologist at Stanford who came to believe that memories were not localised at specific brain sites but were somehow spread out or distributed throughout the brain as a whole.

A decade later, biologist, Paul Pietsch set out to disapprove Pribram's theories. In a series of over 700 operations on salamanders, he discovered that their learned behaviour was not affected by repositioning, reversing, or even 'shuffling' the brain. After recovering from the operation, their behaviour returned to normal. These discoveries perplexed Pribram until he discovered holography. As discussed in Chapter 5, if it was possible for every portion of a piece of holographic film to contain all the information necessary to create a whole image, then it seemed equally possible for every part of the brain to contain all of the information necessary to recall an entire memory![23] In other words, Pribram viewed memory as a holographic 'image', which is stored over the entire neuronal network in our physical brain. This might explain short-term memory that lasts up to two minutes but it does not explain how we remember long-term memories years later. Even people who have recovered from a prolonged coma due to a head injury can still recall memories that were laid down before the accident. There must be some extra facility, either in the space-time domain or another domain, to fully explain long-term memory.

To clarify these concepts, Figure 6.2 illustrates the relationships between the different entities but note that information underlies them all. An analogy may be made with the visible spectrum, where white light initially appears to be a single entity but it can be separated into seven different colours by a prism, each specified by a discrete underlying wavelength (information set) and yet they are all part of the original beam of white light. While we are alive, we see ourselves as a single entity but both consciousness and self-awareness are separate sets of information that are integrated with our physical body, three things in one.

In 2005, *Cosmos* wrote: *"At the quantum level, particles of matter such as electrons can be in two places at once. In fact, this*

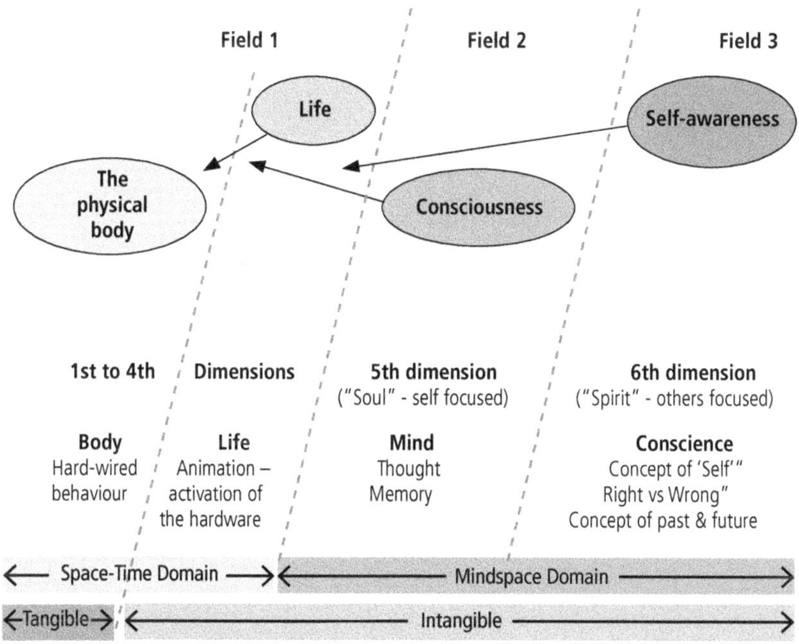

Figure 6.2: Information Sets within the Domain of Life

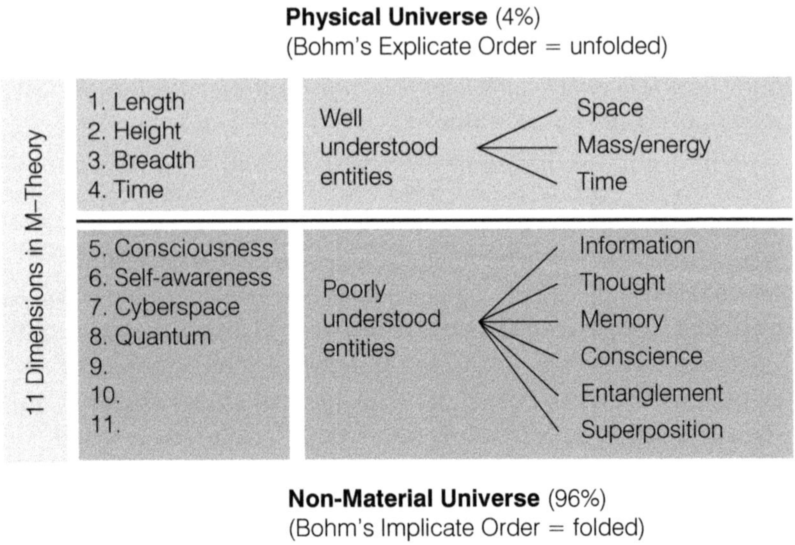

Figure 6.3: The Physical and Non-Material domains within the Universe
As the non-material Universe is 'curled' (according to String Theory) and hence unable to be probed from the physical Universe directly, the dimensions are unknown, so some have been left blank. It seems likely, however, that Dimensions 8 to 11 lie in the Quantum Domain.

seemingly absurd statement is validated by modern electronics. Extrapolated, this means that the building blocks of our world might actually be leading an alternative existence somewhere else. As you sit reading this, there may be another you reading in another parallel world."[24] An alternative interpretation is that you might be appreciating this material from more than one location simultaneously, such as a sixth dimension!

Thus, we have seen that there are many non-material entities within the Universe including: life, consciousness, self-awareness, thoughts, memory, intelligence, language, sleep, mathematics and of course information. Figure 6.3 plots various physical and non-material entities.

An Information Model of the Universe

These ideas may seem strange when first encountered as they include intangibles that have not been included within materialist science to date but Fowler and Kuebler ask: *"Should non-naturalistic explanations be admitted into science? This question is a complete red herring."* They point out that the important issue is not what we define science to be, but whether purely naturalistic explanations are adequate to explain all observed phenomena. If science continues to be restricted to naturalistic explanations, then the issue is whether there are limits to what science can do.[25] Luciano Floridi observes: *"One of the great informational puzzles is how physical signals, transduced by the nervous system, give rise to high-level, semantic* [meaningful] *information…Of course, there is nothing magic in it, but this is no insurance against the fact that the ultimate explanation may one day be astonishing."*[26] After ninety years of research without a coherent explanation, we need to keep an open mind.

Professor of Philosophy, Thomas Nagle went further when he wrote: *"Major scientific advances often require the creation of new concepts, postulating unobservable elements of reality that are needed to explain how natural regularities that initially appear accidental are in fact necessary. The evidence for the existence of such things is precisely that if they existed, they would explain what is otherwise incomprehensible…we should expect theoretical progress in* [the mind-body problem] *to require a major conceptual revolution at least as radical as relativity theory, the introduction of electromagnetic fields into physics – or the original scientific revolution itself…We ourselves are large-scale, complex instances of something both objectively physical from outside and subjectively mental from inside. Perhaps the basis for this identity pervades the world."*[27]

Legitimate scientists have nothing to fear from investigating non-naturalistic phenomena if truth is the goal, as opposed to declaring natural causes necessary by decree.[28] The significant question is: which model generates greater explanatory power?

In an interview in 2009, Richard Dawkins identified four entities with which evolution still struggles: the origin of life, the origin of sex, the origin of consciousness and the rise of morality.[29] However, an Information Model of the Universe that combines both material and non-material elements offers logical answers to all these questions, whereas Neo-Darwinian Evolution does not.

Just as DNA is a physical molecule that acts as a vehicle for the genetic information code and human languages are the vehicles of semantic information,[30] the complex material brain is a vehicle for consciousness and thought, through which we derive meaning. Through the appreciation of information, much of the Universe that remains a mystery to the materialist paradigm becomes comprehensible, but without mind there *is* no meaning.

Chapter 7

Finding Meaning in the Information Ocean

Order and the Universe

Albert Einstein struggled to understand the surprising degree of order within the Universe. He wrote: *"You find it strange that I consider the comprehensibility of the world...as a miracle or as an eternal mystery. Well,...one should expect a chaotic world, which cannot be grasped by the mind in any way...the kind of order created by Newton's theory of gravitation, for example, is wholly different...the success of such a project [as the Cosmos] presupposes a high degree of ordering of the objective world, and this could not be expected...That is the 'miracle' which is being constantly reinforced as our knowledge expands."*[1] Particle physicist, Professor Michio Kaku amplified Einstein's view by saying: *"It turns out that it is no easy matter to create a universe where matter, space and time are stable. For it all to work, the fundamental forces must be unified. And that unification can only take place in higher dimensions...The universe is gorgeous, and it did not have to be that way. It could have been random, lifeless and ugly, but instead, it is full of rich complexity and diversity."*[2] The complexity of the Universe is a surprise because disordered states of matter arise spontaneously whereas ordered states do not.[3] The inescapable conclusion is that the Cosmos is *unexpectedly*, exquisitely organised, a fact that requires explanation.

Information as the Fundamental Entity

So how can we account for this astonishing order in our universe? Paul Davies noted in an editorial in the *New Scientist*: *"The increasing application of the information concept to nature has prompted a curious conjecture. Normally we think of the world as composed of simple, clod-like, material particles, and information as a derived phenomenon attached to special, organised states of matter. But maybe it is the other way around: perhaps the universe is really a frolic of primal information, and material objects a complex secondary manifestation."*[4] This idea was first proposed in 1989 by the eminent physicist John Archibald Wheeler, who said: *"Tomorrow, we will have learned to understand all of physics in the language of information."*[5]

This theme was further developed by Hans Christian von Baeyer in 2001 in a piece entitled: *'In the beginning was the bit'*, where he describes the work of physicist Anton Zeilinger. Zeilinger proposed that quantum mechanics could be understood better if each piece of information relating to an elementary property, like the spin of an electron, was understood to be carrying one bit of information (spin is either 'up' or 'down'). John Lennox comments: *"This proposal that information be regarded as a fundamental quantity has profound implications for our understanding of the universe."*[6] Note that these 'bits' of information are far more fundamental or basic than the 'infobits' discussed in Chapter 1. Professor of Quantum Mechanical-Engineering, Seth Lloyd refers to these as 'quantum bits' or 'qubits'.[7] Infobits are useful for measuring macroscopic entities while atom-sized (called 'Plank scale') information is more appropriately measured in qubits. At this stage the qubit is a useful concept but impossible to use to calculate information content due to the vast complexity of our universe. Yet if quantum mechanics can be measured in qubits of information (that is, zeros and ones), then it is digital, meaning that our universe is digital from its most basic particles to its most complex structures.

Andrew Gosler of Oxford agrees, writing: *"...all that exists, exists* as *information."*[8] Lennox also concludes that information is more fundamental than mass-energy, is non-material and has no mass outside its carriers.[9] Thus, the material Universe is based upon the invisible and non-material entity of information. Material entities are just the visible and measurable expressions of information and its residence within the physical space-time domain. This vast amount of information has been named the 'infosphere'. If the concept still seems strange, Luciano Floridi suggests imagining the whole universe in terms of chemistry. Every entity and process can be described chemically. Human beings, for example, are mostly water but from an informational perspective, the atoms from which they are composed can be described as clusters of data, that is, as informational objects.[10]

Thus information, even though it is often not easily identifiable, is the entity from which everything else is made. However, our universe lacks an internal mechanism to generate this information. Therefore, the information must have originated from somewhere outside our physical Cosmos, a separate domain from whence the information to specify all energy/ matter was imported. As the source of everything in the Universe, this information must have preceded the formation of the Universe. Seth Lloyd concludes: *"Information and energy play complementary roles in the universe: energy makes physical systems do things. Information tells them what to do."* Thus *"...the primary actor in the physical history of the universe is information"*[11] Paradoxically, the most significant determinant of the Universe is one of the least detectable. All physics and chemistry act as the mechanisms or levers by which information manages the Universe. This is a vast research field, the study of which I would suggest should be expanded into a new discipline called: 'Information Mechanics'. The term: 'Information Science' refers to the processing and material vectors of information. Information Science needs to be broadened to look at the wider implications of fundamental or pure information to include its other non-material characteristics.

I am therefore proposing that the current materialist paradigm be replaced by an information paradigm which includes the former but extends it. The material paradigm has been very useful for four centuries but in the same way that the study of mathematics was incomplete without the concept of zero, the study of the Universe is incomplete without the entity of information with its attendant non-material characteristics.

Information and time

Like information, time is another intangible entity and yet it has physical properties with dimensions in the space-time domain and it varies with gravity, mass and acceleration.[12] It has previously been noted that information is outside of time and does not change with time. The physical medium in which information is stored (such as papyrus) may age in our physical space-time domain but the information itself does not. This means that information *cannot* tell you how old it is, such as when it was generated. Thus, non-material information does not help with determining the age of the Universe or the age of life on Earth. That must be determined from its physical carriers.

Materialism and Metaphysics

The study of non-material entities (sometimes referred to as *Metaphysics*), is often considered to be outside the realm of (empirical) science and hence not considered to be 'real', which begs the question: 'What is reality?' Oxford biochemist and theologian, Arthur Peacocke cites the dictum: *"For to be real is to have causal power"*[13] and there is no doubt that information, mathematics, human thought and human consciousness have causal power. Therefore, to summarily dismiss all non-material entities as not being real is invalid as it would be denying our human existence. As a result, some scientists have begun to ask questions.

In 1999 Paul Davies wrote: *"How did stupid atoms spontaneously write their own software? And where did the very peculiar form of information needed to get the first living cell up and running come from. Nobody knows. There is no known law of physics able to create information from nothing."*[14] In an effort to answer this conundrum, Davies speculates that maybe some sort of self-organising physical process could achieve a certain threshold of complexity after which some 'complexity laws' might self-organize and lead to the appearance of life.[15] The problem is that, even if they did exist, the laws he is appealing to would require prior information input too.

From an information perspective, materialism and metaphysics are two parts of an overlapping greater whole, which are *both* necessary to fully understand our universe. Materialists may propose that non-material entities are simply 'emergent properties' of material ones, but it has already been observed that information is independent of its material vehicles and Davies argues that it precedes them. So, in the light of the evidence, logic suggests that both material and non-material entities exist and that *both are 'real'*. Nobel laureate and father of quantum physics, Max Planck stated: *"As a man who has devoted his whole life to the most clearheaded science, to the study of matter, I can tell you as a result of my research about atoms this much: There is no matter as such! All matter originates and exists only by virtue of a force which brings the particles of an atom to vibration and holds this most minute solar system of the atom together...We must assume behind this force the existence of a conscious and intelligent Mind. This Mind is the matrix of all matter."*[16]

So, insisting that every entity in the Universe must be 'material' is like replacing the 'Flat Earth' Principle with a 'Shallow Universe' Model! Once one's eyes are open to non-material entities, one realises that they have always been there; it is just that we have not been attuned to recognising them.

Categorising the information required for life

In Figure 4.2, it was seen that the minimum requirements for the first cell included at least nine different entities that were *simultaneously* required for life. These nine components can now be categorised into four different spheres: matter, energy, information and life itself (see Figure 7.1 on the next page).

Charles Darwin gave an explanation for the existence of life on Earth but as was seen in Chapter 4, the evidence is circumstantial. Most notably, Neo-Darwinian evolution has no explanation for the origin of the information in the first cell. Before there was a first cell, there cannot be any *intrinsic* information so there had to be an *extrinsic* information source which can only be explained by an external biological information generator. The nature of this information generator is yet to be empirically identified but as we will see below, there are some clues.

The Biological Information Hypothesis

Many thousands of scientists have expected that with time and

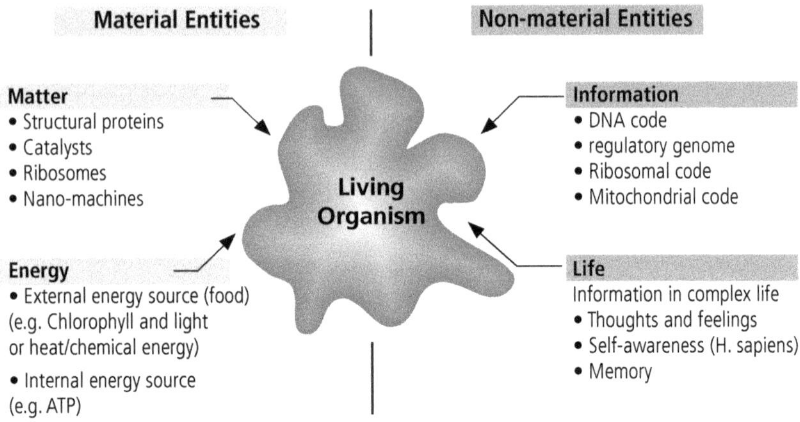

Figure 7.1: Material and Non-Material requirements for Life
Living organisms cross the divide between the tangible and intangible. All life requires a combination of these components to survive, no matter how simple or complex the organism.

increasing technology, the mechanism of random variation with natural selection would be proven to increase complexity in biological systems but microbiological advances have done just the opposite, revealing stunningly complex nano-systems with bewildering arrays of inter-connecting chemical networks, demonstrating that the initial assumption that basic life forms would be 'simple' is just not true. Not only are the simplest life-forms that are presently found on Earth so sophisticated that they cannot be reproduced from scratch (even in a laboratory) but a possible prototype that could have initiated life, cannot be devised, even in theory. (No one has a clue how to build a mitochondrion.) In addition, the places in the Universe where life can survive, once it has begun, are very limited. Most of the Universe is hostile to life. Despite six decades of scouring the Universe, not one other planet has been found to contain any sign of life, let alone be conducive to the spontaneous appearance of intelligent life and the laws of biology suggest that if intelligent life does exist anywhere else in the Universe, it will be exceedingly rare. These prerequisites for life can be summarised in the *Biological Information Hypothesis* which states: information is the grand conductor of life and in all biological systems, information precedes life. This is the concept of biological information primacy.[17]

Characteristics of Information

In summary:

- information is intangible and yet it is real (it has *causal power*);
- information is neither matter nor energy but pre-dates them;
- information is transmissible but…
- information requires a vehicle to be transferred;
- information is independent of time;
- information can be both quantitative and qualitative;
- meaningful information needs to be actively generated (the Golden Rule);
- information can be stored;
- information can be digital (numerical) or analogue (such as text);
- information comes at a cost, both to generate and maintain;
- the existence of information does not necessarily explain its causation.

From this knowledge, two basic principles of information can be defined:

1. **The law of information primacy***: information is the fundamental entity of the Universe which specifies everything, including matter and energy*. Therefore, everything in the Cosmos depends on information which must be generated first.
2. **Information is independent of time.**
   These principles can be combined to build an Information Model of the Universe. (See Figure 7.3 on page 120)

The Origin of Information

Our investigation of information has led us from the smallest cell to the complexity of the human brain and from the non-material, subatomic realm to the vastness of the Cosmos. Underpinning everything is information, which appears to be distributed in a digital, sometimes holographic manner. This has provided explanatory power for poorly understood entities but the origin of the information that we can presently detect is a problem. Walter Bradley, Professor of Engineering, points out that there are no free lunches in nature when one is trying to explain the origin of information[18] while Arno Penzias, Nobel laureate in physics observes: *"Astronomy leads us to a unique event, a universe which was created out of nothing, and delicately balanced to provide exactly the conditions required to support life. In the absence of an absurdly-improbable accident, the observations of modern science seem to suggest an underlying, one might say, supernatural plan."*[19]

The Digital Universe

Even in 1623, as Galileo was beginning to understand the order within the Universe, he recognised that: *"The great book of nature is written in mathematical characters"*[20] and in Chapter 2 we saw that we do indeed live in a universe that can be accurately described mathematically. The Nobel Laureate, Paul Dirac, viewed the world not as made of things, but constituted of an abstract mathematical structure which shows us how things appear and how they behave when we encounter them.[21] Cosmologist, Sean Carroll goes even further, commenting: *"Each state of space…[is] a matrix of numbers, one number for each point in space."*[22] Once again the evidence points to a digital universe with an underlying order. Thus information is the cosmic building block. However, if the Universe is defined by a digital information matrix, which we previously discovered was *"finely-tuned for life"*, the implication is that this meaningful information was not random but pre-specified. This suggests that there was intention behind its formation which begs the controversial question, was it designed?

Design

In information terms, *design* is the intentional arrangement of information. Sometimes the purpose of a system is obscured from an observer who stumbles upon it for the first time and the information (and hence the design) goes unrecognised but usually its purpose can be discerned by examining the system.[23] Positively identifying design is not always that easy though. If something from the past is clearly designed, such as an inscribed clay tablet, it is referred to as an artefact. In this case, the writing on the tablet identifies it as something manufactured rather than being an object made by random processes. So, what happens if no one was there to witness the event but it is still suspected that the object was designed? The hallmark of any design is the intentional introduction of information. The difficult thing is to discern if there was deliberate information input or not. A typical example might be the laws of the Universe; were they random or did some form of pre-specified information influence their formation? From an information perspective, if meaningful information content is identified, whether intrinsic or extrinsic, whether the introduction of the information was observed or not, there is an element of 'design'; whereas if there is no meaningful information input, then the only conclusion is that the events occurred by 'chance'. Time and direct observation can be removed from the equation.

The next point to note is that design requires a designer and chance requires the absence of a designer. As digital computer programs require information arranged in a particular way, we know it is not there by random chance. They are the product of a *program designer*. With computer programs it is easy to discern the design (structured information) within the digital system. But do all things that are digital, imply design? If events occurred by chance, then a limited degree of order may occur randomly but it cannot be expected. When throwing a standard die, three sixes in a row does rarely occur by chance but to throw the number six, one hundred times in a row would make one doubt that it was a standard die. Certainly, the multilayered, digital precision that is observed in our universe appears beyond the realms of blind chance. So since the Universe is specified by precise digital information, does the same logic apply? It would appear that no other conclusion is possible; the choices are shown in Figure 7.2 on the following page.

Paul Davies wrote: *"It is hard to resist the impression that the present structure of the universe, apparently so sensitive to minor alterations in numbers, has been rather carefully thought out...The seemingly miraculous concurrence of these numerical values must remain the most compelling evidence for cosmic design."*[25] Owen Gingerich, Harvard Professor of Astronomy similarly commented:

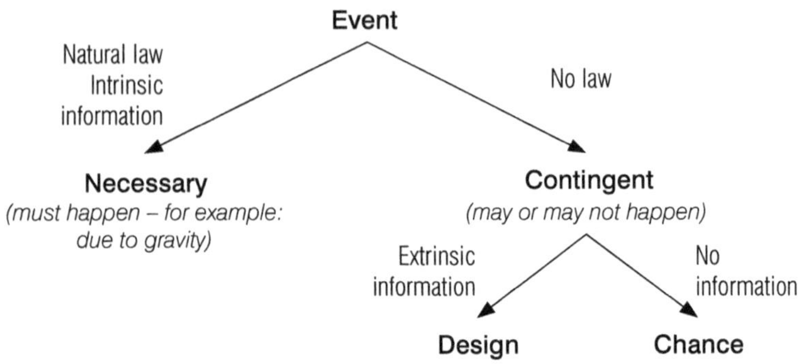

Figure 7.2: Information Outcomes in Our Universe
Note that Natural Law results from previously specified intrinsic information.[24]

"A commonsense and satisfying interpretation of our world suggests the designing hand of a super intelligence."[26]

The other startling implication of a digital universe is that the digital domain would be the primary domain and our familiar space-time realm is a dependent secondary dominion – a projection or manifestation of the veiled, metaphysical, digital realm. Thus, Space-Time is an emergent property of information.[27] This was the conclusion David Bohm came to from his study of quantum physics.[28]

The Information Model

This concludes our information survey. Table 7.1 summarises the key relationships within the infosphere from which an Information Model of the Universe can be constructed. As with computers, there are both platforms and applications. In Chapter 1 we began with the physical carriers of information in the ***Tangible*** domain which is **Domain 1**; this is where we *experience* life in the Cosmos where the information provides physical structure.

Software information running on computer hardware is said to be in ***Cyberspace***. It is a non-material digital processing realm that is instructional and deterministic. It has been discussed for three decades but no one quite knows where or what it is. Although it is 'within' your computer, it is not the same as the microchips which it requires to function in **Domain 2**. It resides in dimension 7 in Figure 6.3 on page 106. ***Mindspace***, which

	1	2	3	4
omains	Tangible	Cyberspace	Mindspace	Quantum
echanics	Einsteinian Relativity	Shannon (syntactic) data management	Meaningful (semantic) info management	Quantum mechanics
perations	Analogue	Digital	Holographic	Vibrational
haracter	Physical (particle/wave)	Granular (deterministic)	Disseminated	Fuzzy/blurry smeared
unction	Structural (platforms)	Instructional processing	Interpretive – mind, free will & meaning	Actualising
cope	Local		Non-local	
nplication	Bounded by time	Non-material & timeless		Depends on interpretation
imensions	1 to 4	7	5 to 6	8 to 11

Table 7.1: Key relationships in the Infosphere

accommodates mind, thought, memory and freewill, is similar to Cyberspace yet vitally different as it demonstrates holographic properties unknown to computers and it operates as an active agent that somehow imparts meaning. This is **Domain 3**. The ***Quantum*** domain remains difficult to understand but impossible to ignore as most modern technology including smart phones and computers depend upon it. It also seems to have a role in giving physical matter its properties by utilising quantum information to actualise everything else – **Domain 4**. In String Theory, the Quantum domain may occupy dimension 8 to 11. (See Figure: 6.3 on page 106) Thus, cyber-processing and mind appear to be applications that run on platforms built from the Tangible and Quantum domains respectively. Although the four domains are recognisable, we can only access the information in the different domains if we have the right vehicle to do so. Some examples are your computer and your ear which are both analogue (physical) to digital interface devices.

> The Quantum domain appears to occupy the region where √-1 is required to make sense of the underlying reality (which is impossible in the material Space-time domain!)[29]

Mapping the information domains

The characteristics of the Information Model in Table 7.1 are mapped in Figure 7.3 to aid our understanding of the relationships. Note that an *Information Fabric* underlies the Space-time fabric that forms our physical cosmos. We previously observed that the information that specifies the quantum domain must have been introduced at the formation of the Universe.

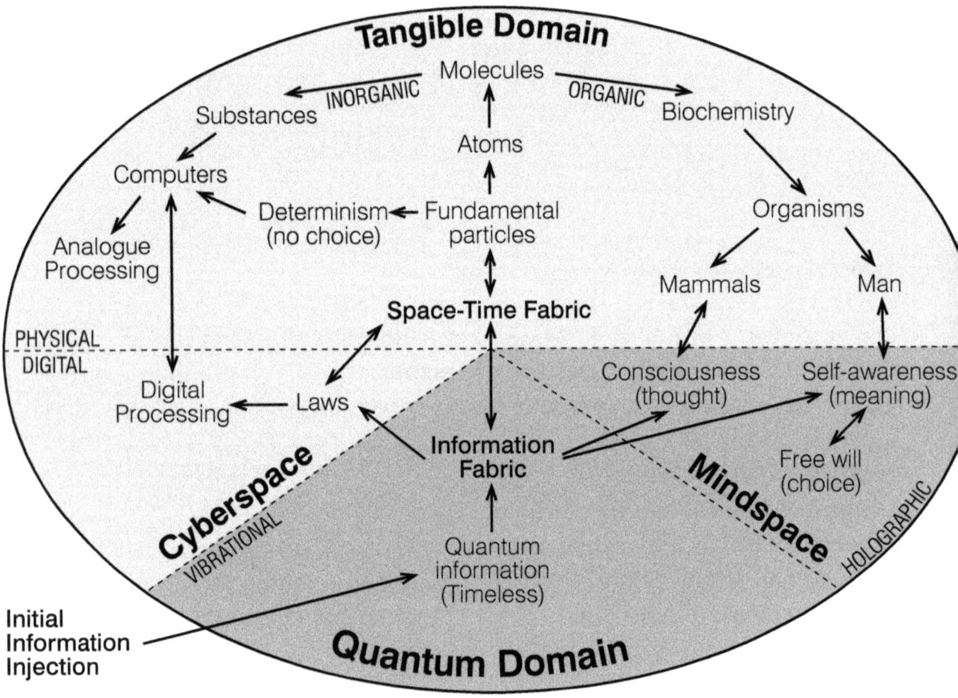

Figure 7.3: Information map of the Universe
Our familiar tangible physical world is shown in the top half of the diagram while the intangible is below. The interfaces between the four domains are poorly understood but the relationships can be inferred from the information that is evident in the Universe.

Thus, we have established that the Universe is highly ordered due to a vast amount of precisely specified digital information and it had a beginning. But where did all this information originate? What does the Standard Model predict was there before the 'hot Big Bang'? Edgar Andrews explains: *"The answer is truly nothing.*

Everything that exists materially is contained within the expanding volume of space itself. There isn't even a vacuum or a void outside of space - just nothing at all."[30] The operative word here though is *"materially"*; information is *non-material*. What if some source of information were outside our material universe that caused it, not ex nihilo ('out of nothing') but by introducing information, in a similar manner to the way that a TV signal creates a 'world' within your TV or DNA brings information to create a new human life? This information injection model avoids the problems of creating an ordered universe out of an uncontrolled explosion and being outside our material world, this information source would not be restricted by the same limitations as previous models, such as the creation of an entire material universe out of nothing, which is required by the Big Bang model. However, if the Big Bang was a 'controlled' event, rather than an uncontrolled one, then there must have been an *information generator* to create this controlling effect.

The Information Generators

Our previous investigations have identified the need for both a biological and a physical information generator but are they the same or different? To answer this question, note that the biological information in DNA, which is essential for all life, requires a physical universe in which to exist, with physical carriers (molecules) to convey the code. The physical information generator must therefore have been active prior to the biological information generator although it is quite possible that a single universal information generator performed both functions. Whatever the process that supplied enough detailed information to specify the Universe, it must have been unimaginably sophisticated. Ockham's razor would suggest that there was just one rather than two information generators.[31]

Ockham's razor: Attributed to the 14th Century English Franciscan thinker, William of Ockham, this is the general principle that one should favour an hypothesis that makes the fewest new assumptions among competing hypotheses that are equal in other respects. It does not (falsely) assert that the simplest explanation is always likely to be the correct one.[31]

The Universal Information Generator

The Law of Information Primacy implies the existence of a universal information generator (UIG). Further, the UIG must have been outside our universe to generate the information to create the Cosmos, so it lies outside time and has no mass and thus is non-material. Therefore, we cannot use materialistic methods to investigate the UIG. As the UIG is outside time, it must be *timeless* so must have had no beginning and will have no end. The options for the identity of the UIG are limited and not new. They are summarised in Figure 7.4.

Precision Thinking

So how do we connect the remaining pieces of information? In Chapter 4 we investigated the clash between Pasteur and Darwin because they both could not be right even though both

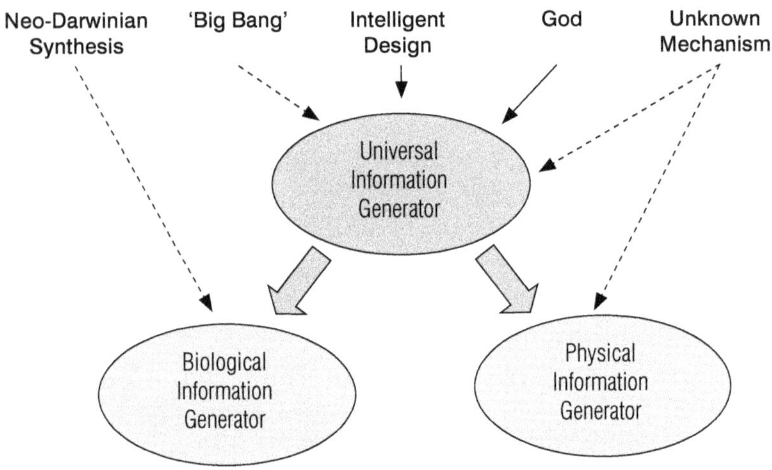

Figure: 7.4: The Formation of the Universe
The number of possible alternatives for the causation of the Universe is small. Chance has not been included as it has already been shown to be inadequate to explain the extraordinary volume of precisely defined information within the Universe. The 'Big Bang' has been included as it is the currently accepted model however it does not provide a mechanism for inducting information, which is a prerequisite for the formation of the Universe. Likewise, the Neo-Darwinian Synthesis also has a problem with information generation (as seen in Chapter 4). It has been included here to demonstrate the limitation of its scope in an information context.

are highly esteemed scientists. George Wald explains: *"We tell this story [regarding Pasteur's experiments disproving spontaneous generation] to beginning students of biology as though it represents a triumph of reason over mysticism. In fact it is very nearly the opposite. The reasonable view was to believe in spontaneous generation; the only alternative, to believe in a single, primary act of supernatural creation. There is no third position."*[32] But where does this leave materialism?

Richard Dawkins in his concluding chapter in *The Greatest Show on Earth* wrote: *"This is a book about evidence, and we have no evidence bearing upon the momentous event that was the start of evolution on this planet. It could have been an event of supreme rarity. It only had to happen once, and as far as we know it did happen only once. ...The beginning had to be simple, and evolution by natural selection is still the only process we know whereby simple beginnings can give rise to complex results."*[33] However, if there is no evidence regarding the beginning of life, then how does he know that the beginning *"had to be simple"*? Dawkins is so convinced that macroevolution is true that he *assumes* that the origin of life was simple, yet he is pinning his hopes on a process for which he admits he has no evidence and no idea of life's origin. Whether this is woolly or wishful thinking, it is *not* science and does not account for the evidence, the information that is so prevalent in our universe. An information model, in which the formation of the Universe is initiated by the introduction of information, provides a much better fit with the data that we can detect and has explanatory power regarding the origin of life, which Dawkins admits the Neo-Darwinian Theory lacks.

If the profoundly complex information in DNA had to be actively generated rather than evolve by chance, then the DNA code was intentional rather than having been an accident and life is not an accident either. We, as human beings, are the result of optimisation in the same way that a building designed by an architect has been optimised. George Wald summarized Richard Dawkins' dilemma when he wrote: *"One only has to contemplate the magnitude of the task to conclude that spontaneous generation of a living organism is impossible"* however he then unexpectedly concluded that: *"Yet we are here - as a result, I believe, of spontaneous generation."*[34] Through this curious logic, Wald manages to uphold both Pasteur, who demonstrated that spontaneous generation does *not* occur and Darwin, who believed that it must! Viewing through an informational lens precludes this type of contradictory thinking, as information (digital code) is precise and specific.

Yet empirical science is supposed to be the study of defined entities using specified methods by precise minds, so how can hard-headed scientists using the same data come to diametrically opposed conclusions for so long? The problem is that science is the study of observations, from which we construct theories to fill the gaps in our knowledge (information that is not available at that time). There are many critical points in constructing an accurate scientific theory but one crucial starting point is in deciding which information is relevant, such that it warrants inclusion, as opposed to that which does not. This is where any preconceived bias comes into play, which can inappropriately weight the value of the evidence we have selected. If the theory turns out to be erroneous but we then prefer the *theory* rather than new (conflicting) observations, we have put the cart before the horse and the resulting predictions may be deceiving. The main problem is that despite all our emphasis on scientific method, we as human beings tend to see what we want to see. Fred Hoyle observed: *"If this were a purely scientific question and not one that touched on the religious problem, I do not believe that any scientist who examined the evidence would fail to draw the inference that the laws of nuclear physics have been deliberately designed..."*[35]

Unfortunately, certain influential sections of the academic community have attempted to arbitrarily restrict science to the study of physical entities alone. These are more easily explored by experimentation but just because something cannot currently be confirmed experimentally, does not *prove* that it does not exist and hence be excluded from study by science. It may just show the limitations of the science of the day. It is clear that non-material entities such as information *cannot* be treated materially, so a new discipline is needed to study it. Hence the term: 'Information Mechanics'. This does not negate those parts of the materialistic, experimental model that are successful. It just points out that the study of information is of equal importance to the study of particle physics and that a different modality is required to study it. In fact, it is my belief that for science to advance, the *parallel* study of both the material and the non-material will be unavoidable, as information underpins the entire Universe. However, the opportunities offered by digital information will not be an instant panacea for everything that is currently unclear or unknown, as information is a tool to aid analysis, not an answer in itself. What is patently clear is that integrating Information Mechanics into mainstream science will provide new insights that may remain concealed without it. As an example, we will revisit the Copernican Principle.

The Copernican Principle

When Copernicus placed the Sun at the centre of the Solar System, there began a long chain of discoveries regarding the Earth's position within the Universe. The realisation that the Earth was not at the centre of the Solar System, let alone at the centre of the galaxy, led some people to question the significance of the Earth and the importance of humankind's existence in 'the grand scheme of things'. If human beings were a special creation of God, wouldn't he have placed them in the most prestigious place, at the centre of the Universe? Instead, Earth appeared, in Carl Sagan's words as: *"a lonely speck in the great enveloping cosmic dark."*[36]

However, humankind's place in the Cosmos is more appropriately assessed by noting that in terms of information, the most complex entity in the Universe is the human brain. Although many have tried to find other intelligent life, none has been found to date. So, far from being an insignificant species on an inconsequential planet, the Earth and especially humankind represent the informational pinnacle of the Universe. New discoveries in the future may change this conclusion but today, based on the *evidence*, this is *fact*.

Sagan's assessment is based on 'geographical' (actually 'cosmographical') information and if you are into real estate, then position is everything. But Gonzalez and Richards have shown that being in the centre of things in universal terms (that is, at the centre of the galaxy) will land you in the middle of a giant black hole![37] However, if we take an information perspective rather than a cosmographical one, humankind has an optimal location from which to observe and gain an understanding of the Universe. By this line of thinking, humankind occupies, not one of the least, but one of the most valuable places in the Cosmos.

Sources of Information

Thus, we live in a precisely defined universe. The degree of precision can be gauged by the error tolerance in the replication of the DNA code or the extreme specificity required for balancing the charge on an electron. This precision reflects the quality of the information generator(s). If information could create itself or there was some naturalistic mechanism that could generate all information, then looking for an external source to reveal the origin of the Universe is probably unnecessary (although it is not impossible that both an external Creator and an internal information generator could exist). An exclusive naturalistic information generator would also imply that the Bible and the other ancient religious books were erroneous because there would be no one from whom that revelation could have come but as there is no naturalistic information generator to account for all the information we observe in our universe, we are forced to conclude that there was an external information source.

Whoever you are, you cannot escape the fact that there was an information generator, the only question is who or what was responsible? Materialists will be searching for materialistic information generators and theists will be invoking the power of God. It depends on the evidence you deem to be accurate and relevant.

Truth

It is now time to return to our question from Chapter 1 which derived from Linus Pauling's definition of science. You will remember that Pauling defined science as: *"the search for the truth"*. This implies that we live in an objective world where it can be discerned that some things are a certain way while others are not.[38] The question is, who decides the criteria and how? If we do not like the idea that a divine Creator has revealed 'ultimate truth' in a book such as the Bible and choose to ignore classical metaphysics, as described by Thomas Aquinas, who also described a methodology for discerning morality and truth,[39] the Information Model offers a third alternative. Truth has a positive predictive value and hence it helps us to anticipate the future. Therefore, an alternative definition for truth could be: *an accurate description of the information content within an entity.* The practical problem with this definition is determining how precisely the description in question matches the information content within the entity that is being considered when there is still so much to learn about our universe. It does define an ideal, however, against which everyday situations might be assessed.

Pragmatically, however, there is a saying which goes: *"The truth won't go away"* and the reason is, that in the end, it is the process or idea that is consistent with the laws of the Universe that ultimately works. Hence this is the solution with which we persist because it delivers the results that help us the most. When we have exhausted the store of everything that does not work, we are left with the ones that do, which we call 'truth'. Reality leads us to truth. An individual may consider from their worldview that truth is relative for them, as they ignore what they do not choose to consider but universal truths are exclusive by definition and are not relative, so do not change as post-modernism might suppose. The reason for this is that the fundamental laws of the Universe are mathematically precise and do not change.

Testing the Boundaries

Religion, faith and spirituality are not usually considered to be comparable to legitimate science. Indeed, the Irish philosopher, Bertrand Russell wrote: *"Religion is something left over from the infancy of our intelligence. It will fade away as we adopt reason and science as our guidelines."*[40] However, after four hundred years of modern science, religion is taking a long time to die. John Lennox takes the opposite view, maintaining that it was the concept behind the Biblical worldview (that we should expect law and order in the Universe) that led to the rise of science.[40] In case you think that this topic has nothing to do with science and philosophy, Paul Davies comments: *"People working at the theoretical frontiers of physics can't avoid coming up against the big questions. The crossover to philosophy and theology are inevitable."*[41] Thus, unless we explore and probe beyond the current boundaries of science, we will be severely hampered in our attempts to make new discoveries.

Even so, this has not stopped some groups attempting to silence debate. The US National Academy of Sciences declared that: *"Science is a way of knowing about the natural world. It is limited to explaining the natural world through natural causes. Science can say nothing about the supernatural. Whether God exists or not is a question about which science is neutral."*[42] However if, for argument's sake, a metaphysical, eternal God is the universal information generator, but science is not yet able to prove this fact, would such a God under the above definition be 'natural'? After all, He would have had to have existed from before the formation of the Universe! Further, would science be able to legitimately investigate this process that formed the Universe or not?

Faith is another contentious issue. It is often thought of as belief or trust in an imaginary being but faith simply fills in the voids where information is still unknown or where there is a lack of incontrovertible proof. So, faith is involved in any belief system where our knowledge or understanding is incomplete, which applies to most aspects of life and the Universe. It applies to belief in the Big Bang Theory and Darwinian Evolution just as much as belief in a transcendent Creator. Thus, as faith bridges the gaps in our information, we often refer to taking a 'leap of faith' as we jump the holes (or chasms) in our knowledge. John Lennox points out that we need to be able to discern the difference between 'blind faith' and 'evidence-based faith'.[43] When we develop a belief system, it depends upon both the evidence we do have (the facts) and the holes in our information (the gaps). The balance between the evidence and the gaps often determines how successful we are when we take our next step. As an example, imagine a one-thousand-metre-long suspension bridge that spans the genetic chasm between reptiles and mammals as they evolved according to Darwin, each fossil (piece of information) representing one plank in the bridge. The gaps in our knowledge represent missing planks or voids. Thus, the success of your next 'leap of faith' (think of it as the distance that your leap needs to carry you to reach the next secure plank in the bridge so that you do not end up at the bottom of the metaphorical ravine) is directly proportional to the evidence (the number of planks) that you have been able to collect, compared with the number of planks that are yet to be discovered. Unfortunately for Darwin, even after 150 years of searching, over ninety percent of the planks are still missing. Hypotheses, theories and laws can serve as towers to which we can attach the end of our 'knowledge bridge'. The higher the tower (the more accurate the hypothesis) the further we are likely to be able to see from the known into the unknown when we take that next step.

Invoking a 'God of the Gaps' is an oft levelled accusation by materialists of theists who propose a non-material solution where an acceptable naturalistic one does not exist, such as the origin of the first cell. The gaps exist because there is either insufficient information or none that is acceptable to the materialist way of thinking. If a material explanation cannot be found despite diligent research over a significant period of time, yet a non-material explanation can provide a solution to fill the gap in our knowledge, then perhaps the non-material alternative should be investigated as both the material and the non-material are complementary and indispensable parts of the Universe. If a Creator God made the Universe, as Figure 7.4 implies, He is not in the gaps in our human knowledge but the exact opposite. As the author, He is behind the information which forms the planks that make up the knowledge bridge; it is just that some of the planks are physical and some are not. I am suggesting that consideration be given to the planks described by non-material entities too.

And then there is the question of spirituality. David Bohm comments: *"In the prevailing philosophy of the Orient, the immeasurable... is regarded as the primary reality."*[44] It is interesting that the West has emphasised the material portion of our existence and the Orient has focused on metaphysical aspects. Both facets are significant and thus each is incomplete if both the material and intangible aspects are not considered together. Avi Loeb, Professor of Science at Harvard wrote in *Nature*: *"Cultivating other approaches avoids stalling progress by investing only in chasing what might turn out to be 'epicycles'.*

After all, the standard model of cosmology is merely a precise account of our ignorance: we do not understand the nature of inflation, dark matter or dark energy. The model has difficulties accounting for the luminous gas and stars that we can see in galaxies, while leaving invisible what we can easily calculate (dark matter and dark energy). This state of affairs is clearly unsatisfactory. The only way to work out whether we are on the wrong path is to encourage competing interpretations of the known data."[45] Perhaps it is fitting that the last word should go to Albert Einstein who stated: *"Everyone who is seriously interested in the pursuit of science becomes convinced that a spirit is manifest in the laws of the universe - a spirit vastly superior to man..."*[46] Einstein is said not to have believed in a personal God but his statement implies a belief in an ultimate source that created the laws of the Universe. You can call this entity the 'Universal Information Generator' or 'God' (or any other name that you consider is more appropriate) but the laws of the Universe certainly did not happen by accident nor materialise out of nothing.

Materialists have not been able to satisfactorily explain the generation of information in the Universe but there is an implicit assumption that it was generated by the laws of the Universe, bit by bit (literally). The generation of information to form the physical Universe is usually implied to be intrinsic to the 'Big Bang' and its aftermath, while the generation of biological information is attributed to the repeated cycles of mutation with natural selection as described by Charles Darwin. However, from the evidence detailed in this book, it has been shown that these explanations are hopelessly inadequate. These scientists must now search for new mechanisms to account for the generation of the meaningful information that is so abundant in our universe. There are two choices: if a materialistic origin for information is invoked, then it must come from the space-time domain in dimensions one to four and will have mass, be influenced by time and therefore have had a beginning and will have an end. In contrast, if a non-material origin for information is invoked, then it must come from an external domain and be outside time and have had no beginning and will have no end. If the Universal Information Generator is metaphysical and outside time, then it is eternal, had no beginning and therefore no cause. If God is that eternal entity, as Christians believe and the Bible claims, then likewise He had no beginning and thus no cause.

Finding Meaning in the Muddle

In *A Brief History of Time*, Stephen Hawking observes: *"The whole history of science has been the gradual realization that events do not happen in an arbitrary manner, but that they reflect a certain underlying order, which may or may not be divinely inspired. … The initial rate of expansion [of the Universe]…would have had to be chosen very precisely for the rate of expansion still to be so close to the critical rate needed to avoid recollapse. This means that the initial state of the universe must have been very carefully chosen indeed if the hot big bang model was correct right back to the beginning of time. It would be very difficult to explain why the universe should have begun in just this way, except as the act of a God who intended to create beings like us."*[47] In contrast to Hawking's approach, Dawkins introduces the *probability* of the existence of God and concludes that on the law of probabilities, the God of the Bible probably does not exist but as Alistair McGrath points out, although we may be highly improbable, we are here. So the issue is not whether God is probable, but whether God is actual.[48]

John Lennox writes: *"One of Richard Dawkins's main* God Delusion *arguments is that, if God created everything, we would have to ask who created God. But the very asking of this question reveals at once that Dawkins has in mind a created God: 'Who* created *God?' Created gods certainly are a delusion…Dawkins's difficulty must be that he cannot believe in something eternal. Why not? Science certainly does not tell us that there is nothing eternal."* Lennox then asks that if Dawkins' question is valid and he believes that the universe created him, then who created his Creator? All such questions must lead to an ultimate reality - for the atheist the ultimate reality is the Universe, and for the theist the ultimate reality is God. Genesis tells us that God is primary and the universe derivative. This worldview is the exact opposite of ancient polytheism and contemporary secularism, both of which assume that matter is primary and everything else, including mind is derivative.[49] Thus, settling these questions about what came first and what is derived, along with the reality of material and non-material entities, will determine your worldview and hence your interpretation of life and the Universe.

Luciano Floridi notes that several philosophers have accepted the principle that the universe might ultimately be non-material or based on a non-material source and neatly summarises the key findings about information when he writes: "*...the physics of information is consistent with two slogans...The first is by Norbert Wiener (1894-1964), the father of cybernetics: 'information is information, not matter or energy. No materialism which does not admit this can survive at the present day.' The other is by John Archibald Wheeler (1911-2008), a very eminent physicist, who coined the expression 'it from bit' to indicate that the ultimate nature of physical reality, the 'it', is informational, comes from the 'bit'. In both cases, physics ends up endorsing an information-based description of nature. The universe is fundamentally composed of data,...patterns or fields of differences, instead of matter or energy, with material objects as a complex secondary manifestation.*"[50]

Earlier in this chapter, Max Planck invited us to search for *"a Mind that is the matrix of all matter"*. Figure 7.4 illustrates that information leads us to only one realistic option, a creator God such as described in the Bible. Anyone who would challenge that it was God who caused this activation of the Universe and animation of life needs to offer another information generator of equally awesome capability!

But the Bible doesn't stop there. The Book of John identifies Jesus, the second part of the Trinity (Triune God), as the author and creator of the Universe and life. Jesus is variously described as *"a stream of living water"* and *"a fountain of life"* and He promises that by accepting the living water that He offers, we will *"never thirst"*, that He will provide the meaning we seek.[51] So information and hence meaning ultimately derive from Jesus Christ and to find ultimate meaning requires us to find and know Him.

Conclusion

We have seen that everything within our universe, including meaning and understanding derive from pure information, which is not the vehicle carrying the message, but an underlying, intangible entity. This journey into information and meaning has been driven by observed facts with the desire to better explain significant deficiencies in our knowledge such as the origins of the Universe and life, consciousness and sleep, thought and mind. A purely materialistic model cannot satisfactorily explain these things (if at all), but Information Mechanics provides significant understanding. However, Information Mechanics is a discipline in its infancy and there is still much to be discovered. I hope this introduction to meaning and the fundamental information that underlies our universe has been clear and transparent.

We are asked by materialists to believe that life, the Universe and everything were created from nothing, by nothing for no reason, which defies logic (and even common sense)! Many also claim that there is no ultimate meaning in the Universe. However, consider instead that although material objects have no intrinsic meaning, in and of themselves, information in a self-aware brain can generate meaning! Based on the evidence, reason suggests that a self-aware information generator manufactured the information in the Universe and that information is its fundamental component.

Our journey into meaning reveals that despite information being one of the least obvious entities in the Universe, it is ubiquitous and that it has defined the Universe since its inception even though we have only recently begun to realise it. However, once recognised and identified, this information can provide meaning that helps us to perceive life and the Universe in a way that we never could before and it promises to usher in a new era in our understanding of science. We live in a material universe that is underpinned by an intangible information fabric. The application of this information provides meaning that affords context, allowing us to prioritise our lives, giving us direction in our search for significance and a vision for the future. The world over, people are searching for happiness, however it is elusive and transitory and many expeditions to find it end in disappointment. Instead, the discovery of meaning leads to purpose that provides satisfaction and fulfilment. Yet this information that reveals meaning is only a gateway to a deeper reality. A century ago Louis Pasteur, one of the greatest biologists of all time, said: *"Little information takes you away from God but more of it takes you to him."* By continuing to follow the information trail back to its source inevitably will lead you to ultimate meaning, *The Mind* in the Matrix!

Appendix 1

Knowing your Life's Purpose

Ravi Zacharias notes that all the great quandries in life relate to one of four issues: origin, meaning, morality and destiny. In this book I have addressed meaning. Experience suggests that you will only find lasting satisfaction in life when you discover your life's purpose. Austrian psychiatrist, Viktor Frankl said: *"Everyone has his own specific vocation or mission in life... a concrete assignment that demands fulfilment. Therein he cannot be replaced... everyone's task is as unique as his specific opportunity."* Every human being is unique and hence there are things in life that you will be better able to do than anyone else. It will bring you joy and fulfilment when you discover these things.

To identify your life's purpose, ask yourself these questions:

1. What makes my life come alive?
What inspires me, draws me in a particular direction or stirs my emotions? What do I long to do or is there something I find effortless so that when I do it, time seems to vanish?

2. Why was I created?
If you agree with the argument developed in this book, that without a 'Mind in the matrix' we would not be here, then you will realise that we are not here by accident. The Bible teaches that God created each person with a unique purpose

and every life is of great worth, so think about your unique mix of abilities and what opportunities present themselves at this time in history. What experiences have you had, both positive and negative and consider the relationships you have with the people around you. These may be signposts to direct your search. Best of all, pray for God's guidance.

3. Do I believe in my potential?
As a child of God, you are his masterpiece (he never makes mistakes) and there is a path laid out before you but we must be bold enough to grab the opportunities with both hands and stay the course. However, being human, we will not always achieve our goals at our first attempt. Fortunately, there are no 'failures' in God's Kingdom. He simply, in the words of Joyce Meyer, *"invites us to go around the mountain one more time"* as we grow through our journey of life. Think of it as the training we all need to achieve our potential. I believe that few of us really think big enough to completely fulfil that potential! So think big.....*very* BIG!!

4. When should I start?
The answer is: 'Now.' Start imagining!

The answers to these questions may take some time to become clear and often the direction our life takes is revealed sequentially. As we master one skill or gain a certain level of understanding, we will be able to comprehend things that we could not see before. Be patient and persistent and with time, you will become aware of the opportunities before you. In the words of Winston Churchill: *"Never, never, never, never....*[add a few more *never's*]*...never, never give up!"*

Acknowledgements

Books are significant undertakings that require a team effort to come to successful fruition. Many people have contributed to the development of the ideas in this book but I particularly wish to thank Dr Ken Meyer for his input in many areas and especially the quantum mechanics. Also Rev Gary Ezzo for the refining of ideas and definitions and Professor Reuben Rose for improving the development of the argument. Professor Charlie Massy has been a constant encourager.

I am very grateful to Dawn, my wife, who has patiently listened to the arguments and continually gives me feedback on ideas and Tim for his analysis and proof reading.

Finally, I wish to acknowledge the author of the source code, *"the author and perfector of our faith"*, Jesus Christ, who is the alpha and omega, the Mind in the Matrix who provides the Bible to instruct us and guide us through life.

Notes

Foreword

1. Ecclesiastes 1:2 (*New International Version*)
2. Seligman, MEP: In Buie, J (1988) 'Me' decades generate depression: individualism erodes commitment to others. APA Monitor: 19,18. *"People born after 1945 were ten times more likely to suffer from depression than people born 50 years earlier."* https://www.clinical-depression.co.uk/dlp/depression-information/major-depression-facts/ [accessed 3 Aug. 2019]
3. https://www.missionaustralia.com.au/what-we-do/research-impact-policy-advocacy/youth-survey [accessed 29 Nov 2018]
4. Sheppard, RZ: Rock Candy, *Time Magazine* (12 Apr 1971)
5. Floridi, L: *Information: A Very Short Introduction.* Oxford Univ. Press (2010) p.2
6. Davies, P: *The Mind of God.* Penguin Books (2008) p.16
7. Floridi, L: *Information: A Very Short Introduction.* (2010) p.12
8. Davies, P: *The Mind of God.* Penguin Books (2008) p.84
9. Hoyle, F: The Universe: Past and Present Reflections. *Engineering and Science* Vol. 45:2 (Nov 1981) pp.8–12
10. Handy, C: The Seductions of the Infosphere. *Harvard Business Review* 15 Jul 2015 https://hbr.org/2015/07/the-seductions-of-the-infosphere [accessed 25 May 2017]
11. Trigg, R: *Beyond Matter. Why Science Needs Metaphysics.* Templeton Press (2015) pp.52–53
12. McGrath, A: *Enriching our Vision of Reality – Theology and the natural sciences in dialogue*. SPCK, London (2016) p.20

Chapter 1. Searching for Meaning in the Information Age

1. McGrath, A: *The Great Mystery. Science, God & the Human Quest for Meaning.* Hodder & Stoughton UK (2017) p.58
2. Lennox, JC: *Against the Flow* Monarch Books, Oxford. (2015) p.52
3. Battersby, S: The Unseen Agent. *New Scientist* (3073) 14 May 2016 p.28
4. Davies, P & Gregersen, NH (Eds.): *Information and the Nature of Reality. From Physics to Metaphysics*. Cambridge Univ. Press, Cambridge (2010) p.147
5. Lennox, JC: *God's Undertaker. Has Science Buried God?* Lion Books, Oxford (2007) pp.141–2
6. Floridi, L: *Information: A Very Short Introduction.* (2010) p.38
7. *Ibid.* p.9
8. Alberts, B: The Cell as a Collection of Protein Machines. *Cell* 92:291 (1998) *in* Strobel, L: *The Case for a Creator.* Zondervan, Grand Rapids (2004) p.193
9. Lennox, JC: *God's Undertaker. Has Science Buried God?* (2007) p.168

10. Three types of complexity are discussed in Lennox, JC: (2007) pp.145–6
11. Hawking, S & Mlodinow, L: *The Grand Design.* (2010) p.32
12. Lennox, JC: *God's Undertaker. Has Science Buried God?* (2nd Ed.) Lion Books, Oxford (2009) p.63
13. Johnson, P: *Wedge of Truth.* InterVarsity Press, Illinois (2000) pp.53–4
14. Dawkins, R & Lennox, JC: *The Great Debate, Oxford.* (DVD) Fixed Point Foundation, USA (2008)
15. Pauling, L: *No More War!* Dodd, Mead & Co., New York (1958) p.209 *in* Strobel, L: *The Case for a Creator.* (2004) p.28
16. Fowler, TB & Kuebler, D: *The Evolution Controversy.* (2007) pp.124–5
17. *Ibid.* p.128
18. Floridi, L: *Information: A Very Short Introduction.* (2010) p.23
19. *Ibid.* pp.44–45 &2
20. The Rosetta Stone records a decree from ancient Egypt that is written in three different scripts which facilitated the deciphering of hieroglyphics.
21. Sperry, R: *from* Dawkins, R & Lennox, JC: *The God Delusion Debate.* (DVD) Fixed Point, USA (2007)
22. Floridi, L: *Information: A Very Short Introduction.* (2010) p.89

Chapter 2. Our Mathematical Universe

1. Dolphin, L: *Lord of Time and Space.* Good News Publishers, Westchester, Illinois (1974) pp.8–9
2. Ross, H: *The Creator and the Cosmos.* NavPress, Colorado Springs (1993) at alta3b.files.wordpress.com/2016/02/creator-and-the-cosmos1.pdf [accessed 3 Jul 16]
3. Gonzalez, G & Richards, JW: *The Privileged Planet.* Regnery Publishing, Washington DC (2004) pp.174–5
4. Al-Khalili, J: *Black Holes, Worm Holes & Time Machines.* Taylor & Francis Group, NY (1999) p.58
5. *Ibid.* pp.52–3
6. Gaensler, B: Extreme Speed. *Cosmos* 41:71 (2011)
7. Dolphin, L: *Lord of Time and Space.* (1974) p.34
8. Rolston, H: Care on Earth: generating informed concern. *in* Davies, P & Gregersen, NH: *Information and the Nature of Reality: from Physics to Metaphysics.* (2010) p.208
9. Hawking, S & Mlodinow, L: *The Grand Design.* (2010) pp.103–4
10. Gonzalez, G & Richards, JW: *The Privileged Planet.* (2004) pp.197–210
11. *Cosmos* [accessed 5 Oct 16]
12. Lennox, JC: *God's Undertaker. Has Science Buried God?* (2007) p.60
13. Davies, P: *Universe from bit. in* Davies, P and Gregersen, NH (eds.): *Information and the Nature of Reality.* (2010) p.75
14. Andrews, E: *Who Made God? Searching for a Theory of Everything.* EP Books, Darlington, Eng. (2009) p.30
15. Missler,C: *Cosmic Codes: Metacodes.* (CD) Koinonia House (2004)
16. Hawking, S & Mlodinow, L: *The Grand Design.* (2010) p.159–162
17. Lennox, JC: *God's Undertaker. Has Science Buried God?* (2007) p.69
18. Strobel, L: *The Case for a Creator.* (2004) pp.133–4
19. Andrews, E: *Who Made God? Searching for a Theory of Everything.* (2009) p.138
20. *Ibid.* pp. 141–2
21. Davies, P: *The Cosmic Blueprint: New Discoveries in Nature's Creative Ability to Order the Universe.* Touchstone Books, NY (1989) p.203 *in* Gonzalez, G.& Richards, JW: *The Privileged Planet.* (2004) p.195
22. Andrews, E: *Who Made God? Searching for a Theory of Everything.* (2009) pp.142–3
23. *Ibid.* p.148
24. Stenger, VJ: *God, the failed hypothesis: How science shows that God does not exist.* Prometheus Books, NY (2007) p.131 *in* Andrews, E: *Who Made God?* (2009) p.150
25. Andrews, E: *Who Made God? Searching for a Theory of Everything.* (2009) pp.150–2
26. McMullin, E: *From matter to materialism... and (almost) back. in* Davies, P and Gregersen, NH (eds.): (2010) p.34
27. Atkins, P: *Creation Revisited.* Harmondsworth, Penguin (1994) p.143 *in* Lennox, J.C.: (2007) p.62
28. Strobel, L: *The Case for a Creator.* (2004) pp.97–109

29. Davies, P reported by Clive Cookson: Scientists who glimpsed God. *Financial Times*, Apr 29, (1995) p.20 *in* Lennox, JC: (2007) p.63
30. Dawkins, R: BBC Christmas Lectures Study Guide, London, BBC (1991) *in* Lennox, JC: (2007) p.55
31. Lennox, JC: *God's Undertaker. Has Science Buried God?* (2007) p.64
32. *Ibid.* p.67
33. *Ibid.* p.183, Ref. 31
34. Craig, WL: *Reasonable Faith.* (Rev. Ed.) Crossway, Wheaton, Il (1994) *in* Strobel, L & Vogel, J: *The Case for a Creator.* (2004) p.15
35. Al-Khalili, J: Black Holes, Worm Holes & Time Machines. (1999) p.212
36. Hoyle, F: The Universe: Past and Present Reflections. *Engineering and Science* (Nov 1981) pp.8–12 *in* Sarfati, J: (2008) p.238
37. Ward, K: *Why There Almost Certainly Is A God.* Lion Hudson, Oxford (2008) pp.67–70
38. Strobel, L: *The Case for a Creator.* (2004) p.141
39. Hawking, S & Mlodinow, L: *The Grand Design.* (2010) pp.180–1
40. Davies, P: Life, the universe and everything. *Cosmos* 14 (2007) *at* http://www.cosmosmagazine.com/
41. Davies, P: Bit before it? *New Scientist* 161(2171):3 (1999) *in* Lennox, JC: (2007) p.167
42. Küppers, B-O: *Information and communication in living matter. in* Davies, P & Gregersen, NH: *Information and the Nature of Reality.* (2010) pp.180–1
43. Andrews, E: *Who Made God? Searching for a Theory of Everything.* (2009) p.93
44. Rovelli, C: *Reality Is Not What it Seems – The Journey to Quantum Gravity.* Allen Lane, UK (2016) p.28
45. Wigner, E: The Unreasonable Effectiveness of Mathematics in the Natural Sciences. *in Symmetries and Reflections.* Indiana Univ. Press (1967) pp.222–237 *in* Gonzalez, G & Richards, JW: (2004) p.212
46. Moskowitz, C: Tangled Up in Spacetime. *Scientific American* 26 Oct 2016 https://www.scientificamerican.com/article/tangled-up-in-spacetime/ [Accessed 30 Jul 17]
47. Shariatmadari, H: *What is Reality?* Horizon, BBC Productions, UK (2011)
48. Al-Khalili, J: *Black Holes, Worm Holes & Time Machines.* (1999) pp.68–70
49. Davies, P: *The Mind of God.* Penguin Books (2008) p.84
50. Hawking, S & Mlodinow, L: *The Grand Design.* (2010) p.161
51. Drake, S: Discoveries and Opinions of Galileo. Doubleday-Anchor, NY (1957) *in* Davies, P: Universe from Bit. pp.65–91 *in* Davies, P & Gregersen, NH (eds.): *Information and the Nature of Reality.* (2010) p.67
52. O'Connell, C: The Mystery of the Missing Antimatter. *Cosmos* 69 (2016) p.36

Chapter 3. Is there Meaning in the Universe?

1. Rees, M: *Just Six Numbers: The Deep Forces that Shape the Universe.* Weidenfeld & Nicolson, London (1999)
2. Gonzalez, G & Richards, JW: *The Privileged Planet.* (2004) p.136
3. Clark, S: Unknown Earth: How come Earth got all the good stuff? *New Scientist,* 199(2675): 29–30 (2008)
4. Davies, P: Life, the universe and everything. *Cosmos* 14 (2007)
5. Gonzalez, G & Richards, JW: *The Privileged Planet.* (2004) pp.33–4
6. Hawking, S & Mlodinow, L: *The Grand Design.* (2010) pp.152–3
7. Gonzalez, G & Richards, JW: *The Privileged Planet.* (DVD) Illustra Media (2004) – adapted (2019)
8. Gonzalez, G & Richards, JW: *The Privileged Planet.* (2004) pp.197–210
9. Ross, H: *The Creator and the Cosmos.* NavPress, Colorado Springs (1993)
10. For more detail see: Wiles, RB: *The Information Paradox.* Information Press (2014) pp.202–4
11. Missler, C: *Cosmic Codes: Metacodes.* (CD) Koinonia House (2004)
12. Missler, C: *The Beyond Collection: Beyond Coincidence.* (DVD) Ver. 2.0. Koinonia House (2010)
13. Gonzalez, G & Richards, JW: *The Privileged Planet.* (2004) pp.66–7; pp.127–132; p.182; p.253
14. Seiglie, M: Prove there is a God even without the Bible. *The Good News* 17(2):15 (2012)
15. Strobel, L: *The Case for a Creator.* (2004) pp.179–180

16. Dolphin, L: *Lord of Time and Space.* (1974) pp.35–6
17. Bradley, WL: Nature. Designed or Designoid *in Mere Creation* Dembski, WA (ed): *Mere Creation. Science, Faith and Intelligent Design.* InterVarsity Press (1998) p.40
18. Strobel, L & Vogel, J: *The Case for a Creator.* (2004) p.50
19. Davies, P: *The Cosmic Blueprint.* Simon and Schuster, NY (1988) p.203
20. Wald, G: The origin of life. *Scientific American* 191:53 (Aug. 1954)

Chapter 4. Searching for the Meaning of Life

1. Küppers, B-O: *Information and the Origin of Life.* Cambridge, MA: the MIT Press (1990) *in Information and Communication in Living Matter, in* Davies, P & Gregersen, NH (eds.): (2010) p.170
2. Huxley, TH: *Autobiography and Selected Essays.* (Ed. Snell, ALF) Riverside College Classics (2009) *in J. of Creation* 23:3 (2009) p.125 [www.scribd.com/.../Autobiography-and-Selected-Essays-by-Huxley-Thomas-Henry-1825-1895]
3. Venter, C et al: Creation of a Bacterial Cell Controlled by a Chemically Synthesized Genome. *Science* 329 (5987):52–56 (2010)
4. Dawkins, R: *The Blind Watchmaker.* Longman, London (1986) p.112 *in* Lennox, JC: *God's Undertaker. Has Science Buried God?* (2007) p.127
5. *Shorter Oxford English Dictionary.* Oxford University Press, London (1986)
6. Fowler, TB & Kuebler, D: *The Evolution Controversy.* (2007) p.26
7. Darwin, C: *The Origin of Species.* (1859) reprinted by Penguin Classics (1985) p.433
8. Johnson, P: *Wedge of Truth.* InterVarsity Press. Illinois (2000) p.48
9. This is the current name for Darwin's Theory since it has evolved to incorporate new discoveries.
10. Fowler, TB & Kuebler, D: *The Evolution Controversy.* (2007) p.344 (Emphasis in the original)
11. *Ibid.* p.345
12. Lovett, RA: Darwin's Evidence for Evolution. *Cosmos* 24 (2008) p.24
13. Fowler, TB & Kuebler, D: *The Evolution Controversy.* (2007) p.58
14. Marsupials and monotremes (the platypus and echidna) became extinct where they had to compete with placental mammals but survived where they were protected by isolation, as in Australia.
15. Cohen, J & Stewart, I: *The Collapse of Chaos – Discovering Simplicity in a Complex World.* Penguin Books (1994) p.113
16. Linton, A: Scant search for the maker. *Times Higher Education Supplement.* (20 Apr 2001) p.29 *in* Meyer, SC et al: *Explore Evolution.* Hill House, London (2009) pp.104–5
17. Dawkins, R: *The Greatest Show on Earth.* Bantam Press, London (2009) pp117–130
18. *Ibid.* p.126
19. Sarfati, J: *The Greatest hoax on Earth? Refuting Dawkins on Evolution.* Creation Books. Atlanta, Georgia (2010) pp.66–7
20. Shapiro, J: *A Third Way.* (alternative to Darwinism & Creationism). *Boston Review* 22(1):32–33 (1997) p.13 *in* Lennox, JC: *God's Undertaker. Has Science Buried God?* (2007) p.134
21. Behe, MJ: *Darwin's Black Box:* Free Press, NY (2006) pp.185-6
22. Lennox, JC: *God's Undertaker. Has Science Buried God?* (2007) p.119
23. Denton, M: *Evolution: A Theory in Crisis.* Adler & Adler, Chevy Chase, Maryland (1986) p.293
24. Lennox, JC: *God's Undertaker.* 2nd Ed (2009) p.109
25. Dawkins, R: *The Blind Watchmaker* WW Norton, NY (1996) pp.1-3 *in* Johnson, P: *Wedge of Truth.* (2000) p.126
26. Lennox, JC: *God's Undertaker. Has Science Buried God?* (2007) p.117
27. Smith, C: *Codes and Creation.* (CD) Creation Ministries International, Brisbane, Australia
28. Garrett, RA: Mechanics of the Ribosome. *Nature* 400(6747):811–2 (1999) *in* Sarfati, J: (2008) p.167
29. Sarfati, J: *By Design.* Creation Books. (2008) pp.166–7
30. Lennox, JC: *God's Undertaker. Has Science Buried God?* (2007) p.147
31. Yockey, HP: A calculation of the Probability of Spontaneous Biogenesis by Information Theory. *J. Theor. Biol.* 67:377–398 (1977) *in* Sarfati, J: *By Design.* Creation Books (2008) p.158

32. Zouni, A et al: Crystal structure of photosystem II from *Synechococcus* elongatus at 3.8Å resolution *Nature* 409(6821):739–43 (2001) *in* Sarfati, J: *By Design.* Creation Books (2008) pp.126–7
33. Williams, A: Mutations: evolution's engine becomes evolution's end! *J.of Creation* 22(2):61 (2008)
34. Cai, D, Verhey, KJ & Meyhöfer, E: Tracking single Kinesin molecules in the cytoplasm of mammalian cells. *Biophysics Journal* 92(12):4137–44 (2007)
35. Dawkins, R: *The Greatest Show on Earth.* (2009) p.238
36. Wolfenden, R: *Without enzymes, biological reaction essential to life takes 2.3 billion years.* Univ. of North Carolina website *at* http://www.med.unc.edu/biochem/news/without-enzyme-biological-reaction-essential-to-life-takes-2-3-billion-years-unc-study/
37. Popper, K: *Scientific Reduction and the Essential Incompleteness of All Science, in* Ayala, FJ & Dobzhansky, T (Ed.): *Studies in the Philosophy of Biology: Reduction and Related Problems* University of California Press, Berkeley (1974) p.270 *in* Sarfati, J: (2008) p.165
38. Lennox, JC: *God's Undertaker. Has Science Buried God?* (2007) p.131
39. Missler, C: *Cosmic Codes.* Koinonia House (2004) pp.322–4
40. Johnson, P: *Wedge of Truth.* InterVarsity Press, Illinois (2000) p.127
41. *Ibid.* p.131
42. Sarfati, J: *By Design.* Creation Books (2008) pp.216–7
43. Matic, J: DNA Unravelled, *in The Boston Globe* (24 Sep 2007)
44. Putnam NH, et al: Sea anemone genome reveals ancestral eumetazoan gene repertoire and genomic organization. *Science* 317(5834):86–94 (2007)
45. Carroll, SB: *Endless Forms Most Beautiful: The New Science of Evo Devo.* WW Norton & Co, NY (2005) p.139, Ref. 4 *in Creation* 31(4):44 (2009)
46. *Ibid. in Creation* 31(4):44 (2009) p.144
47. Fowler, TB & Kuebler, D: *The Evolution Controversy.* (2007) p.312
48. Prigogine, I & Stengers, I: *Order out of Chaos.* University of Michigan. Bantam Books (1984)
49. Meyer, S: *The Return of the God Hypothesis.* Discovery Institute, Seattle (1998) p.37 *in* Lennox, JC: (2007) p.125
50. In this context, to be 'alive' can be thought of as the ability to reproduce (see Chapter 5)
51. Meyer, SC: DNA and other Designs. *First Things* 102 (2000)
52. Crick, FHC: *Life Itself: Its Origin and Nature.* Simon & Schuster, NY (1981) p.88
53. Britt, R: Panspermia Q and A: leading proponent Chandra Wickramasinghe, (27 Oct 2000) *in* Matthews, M: Life Discovered in Space. *at* http://creation.com/life-discovered-in-space
54. Haldane, JBS: *Data needed for a blueprint of the first organism.* published posthumously *in* Fox, SW (Ed.): *The origins of prebiological systems and of their molecular matrices.* Academic Press, NY (1965) p.12 *in* Andrews, E: *Who Made God?* (2009) p.174

Chapter 5. The Matter of Consciousness

1. Küppers, B-O: *Information and communication in living matter, in* Davies, P & Gregersen, NH: *Information and the Nature of Reality.* (2010) p.177
2. Ward, K: *God as the ultimate informational principle, in* Davies, P & Gregersen, NH: (2010) p.290
3. Russell, B: *History of Western Philosophy.* Routledge, London (2000) p.13 *in* Lennox, JC: *God's Undertaker. Has Science Buried God?* (2007) p.41
4. Briggs, A: And Information Became Physical. *in* Berry, RJ (Ed.): *Real Scientists Real Faith.* Monarch Books, Oxford (2009) pp.77 & 80
5. Frank, A: Minding Matter. *Aeon* 13 Mar 2017 https://aeon.co/essays/materialism-alone-cannot-explain-the-riddle-of-consciousness [Accessed 4 May 2017]
6. Bohr, N: *at* Jones, AZ: About.com Education: Physics – Niels Bohr Quotes http://physics.about.com/od/nielsbohr/tp/Niels-Bohr-Quotes.htm
7. Wieland, C & Farmer, M: *Clash over Origins – Creation vs Evolution Debate.* (DVD) Creation Ministries International
8. Grant, S: From little things, big things grow. *Univ. Sydney Alumni Mag.* Autumn 2009 pp.14–17

9. Encyclopaedia of Mental Disorders (2009) http://www.minddisorders.com/Kau-Nu/Neurotransmitters.html
10. Johnson, P: *Wedge of Truth.* InterVarsity Press, Illinois (2000) p.118
11. Minsky, ML: *The Society of Mind.* Simon & Schuster, NY (1986) p.287
12. Kruszelnicki, KS: *Great Myth Conceptions*. Harper Collins Australia (2004) pp.160–2
13. Missler, C: *Cosmic Codes.* Koinonia House (2004) p.330
14. de Valois, RL & KK: Spatial Vision. *Annual Review of Psychology* 31:309–341 (1980)
15. Floridi, L: *Information: A Very Short Introduction.* (2010) p.11
16. Finkel, E: Fish out of Water. *Cosmos* 2:77 (2005)
17. Wiles, RB: *The Information Paradox.* Information Press (2014) pp.134–6
18. *Ibid.* p.291
19. Missler, C: *Cosmic Codes - Metacodes.* (CD) Koinonia House (2004)
20. Missler, C: *Cosmic Codes.* Koinonia House (2004) pp.339–41
21. https://blog.frontiersin.org/2017/06/12/blue-brain-team-discovers-a-multi-dimensional-universe-in-brain-networks/ [Accessed 24 Jun 2017]
22. Reimann MW, Nolte M, Scolamiero M, Turner K, Perin R, Chindemi G, Dłotko P, Levi R, Hess K and Markram H (2017) Cliques of Neurons Bound into Cavities Provide a Missing Link between Structure and Function. *Front. Comput. Neurosci.* 11:48
23. Andrews, E: *Who Made God? Searching for a Theory of Everything.* (2009) p.203
24. Dawkins, R: *The Greatest Show on Earth.* (2009) pp.405–6
25. *Ibid.* pp.407-8
26. Kauffman, S: *Investigations.* Oxford Univ. Press (2000) p.47 *in* Fowler, TB & Kuebler, D: *The Evolution Controversy.* (2007) p.317 footnote 77
27. Dawkins, R: *The Greatest Show on Earth.* (2009) p.303
28. Wertheim, M: Alas Poor Zombie. *Cosmos* 11:86–91 (2006)
29. Gosler, AG: Surprise and the Value of Life, *in* Berry, RJ (Ed.): (2009) p.187
30. Some maintain there is no 'free will', in which case 'self-will' might be an alternative expression.
31. Eadie, M: A Molecular Basis for Spiritual Experience? *Luke's Journal* 11(2):6 (2006)
32. Gijsbers, A: The Dialogue Between Neuroscience and Theology. *Luke's Journal* 11(2):8–9 (2006)
33. Hawking, S & Mlodinow, L: *The Grand Design.* (2010) pp.30–4
34. Talbot, M: *The universe as a hologram: Does objective reality exist... or is the universe a phantasm?* Harper Collins, NY (1991)

Chapter 6. Mind and Meaning

1. de Duve, C: *Life Evolving.* Oxford University Press, NY (2002) p.199 *in* Lennox, JC: *God's Undertaker. Has Science Buried God?* (2007) p.74
2. Andrews, E: *Who Made God? Searching for a Theory of Everything.* (2009) pp.249–251
3. *The Collected Papers of Bertrand Russell*, Vol. 12: Contemplation and Action, 1902–1914 Routledge, London (1985) *in* Andrews, E: (2009) pp.251-2
4. Andrews, E: *Who Made God? Searching for a Theory of Everything.* (2009) pp.252–3
5. Brooks, R: Professor of Evolutionary Biology Univ. of NSW, Sydney. *Sydney Morning Herald*: News Review (Sep 24–25, 2011) p.18
6. Profoundly mentally impaired people appear to lack self-awareness, which suggests that a fully competent hardware platform is necessary for self-consciousness to operate.
7. Dunbar, R: Why are Humans not Just great Apes? *in What makes us human?* (ed) Pasternak, C: Oneworld Publications, Oxford (2007) pp.37–48 *in Cosmos* 20:47–9 (2008)
8. Davila Ross, M, Orwen, MJ & Zimmermann, E: Reconstructing the Evolution of Laughter in Great Apes and Humans. *Current Biology* 19(13):1106–11 (2009)
9. Morelle, R: *Scientists tickle animals to find laughter clues.* BBC News 30 Dec 2011 *at* http://www.bbc.co.uk/news/science-environment-15880045
10. Brain-controlled wheelchair. *Australian Medicine* Aust. Med. Assoc. (21 Sep 2009) p.20
11. Greenfield, S: *Brain Story.* BBC video series (2001)
12. Thuesen Bleeg, M: Cat naps. *Cosmos* 34:46–7
13. Cirelli, C & Bushey D: Sleep and Wakefulness in Drosophila melanogaster. Molecular & Biophysical Mechanisms of Arousal, Alertness & Attention. *Annals of the NY Acad. of Sciences* 1129: 323–9 (2008)

14. Fanjul-Moles, ML et al: Circadian Modulation of Crustacean Hyperglycaemic Hormone in Crayfish Eyestalk and Retina. *Chronobiology International* 27(1):34–51 (2010)
15. Deacon, TW: *The Symbolic Species: The Co-Evolution of Language and the Brain.* Norton, NY (1997) p.21 *in* Rolston, H III: Care on Earth: generating informed concern. *in* Davies, P & Gregersen, NH: (2010) p.226
16. Moreland, JP: *in* Strobel, L: *The Case for a Creator.* (2004) pp.262–3
17. Haselhurst, G: *Essay on Life & Ideas of David Bohm (1917–1992)* p.9 http://www.spaceandmotion.com/Physics-David-Bohm-Holographic-Universe.htm. See: *Holographic Universe* http://www.essays.cc/free_essays/e4/dkt106.shtml
18. Missler, C: *Cosmic Codes.* Koinonia House (2004) p.330
19. Johnson, PE: *Wedge of Truth.* InterVarsity Press, Illinois (2000) p.123
20. Penfield, W: *The Mystery of the Mind: a Critical Study of Consciousness and the Human Brain.* Princeton Univ. Press, NJ (1975) pp.76–78 *in* Strobel, L: *The Case for a Creator.* (2004) p.258
21. Moreland, JP: Interview *in* Strobel, L: *The Case for a Creator.* (2004) p.258
22. Haselhurst, G: *Essay on Life & Ideas of David Bohm (1917–1992)* pp.7-8 http://www.spaceandmotion.com/Physics-David-Bohm-Holographic-Universe.htm
23. Missler, C: *Cosmic Codes.* Koinonia House (2004) pp.329–330 – See Paul Pietsch, *Shufflebrain: the Quest for the Holographic Mind*, Houghton Mifflin, Boston, MA (1981)
24. Finkel, E: Fish out of Water. *Cosmos* 2: 77 (2005)
25. Fowler, TB & Kuebler, D: *The Evolution Controversy.* (2007) p.356
26. Floridi, L: *Information: A Very Short Introduction.* (2010) p.86
27. Nagle, T: *Mind and Cosmos. Why the Materialist Neo-Darwinian Conception of Nature is Almost Certainly False.* Oxford Univ. Press (2012) p.42
28. Fowler, TB & Kuebler, D: *The Evolution Controversy.* (2007) p.357
29. Boyle, A: The not-so angry evolutionist. http://cosmiclog.msnbc.msn.com 14 Oct 2009 *in* Sarfati, J: (2010) p.27
30. Ward, K: *God as the ultimate informational principle, in* Davies, P & Gregersen, NH (Eds): *Information and the Nature of Reality.* (2010) p.296

Chapter 7. Finding Meaning in the Information Ocean

1. Einstein, A: Letters to Solovine: 1906-1955. Philosophical Library, NY (1987) p.131 *in* Lennox, JC: *God's Undertaker. Has Science Buried God?* (2007) p.59
2. Finkel, E: Fish out of Water. *Cosmos* 2:78–9 (2005)
3. Andrews, E: *Who Made God? Searching for a Theory of Everything.* (2009) p.22
4. Davies, P: Bit before it? *New Scientist* 161(2171):3 (1999)
5. Lennox, JC: (2007) *God's Undertaker. Has Science Buried God?* p.167
6. von Baeyer, HC: In the Beginning was the Bit. *New Scientist* 169(2278):26 (2001) *in* Lennox, JC: (2007) p.167
7. Davies, P & Gregersen, NH: *Information and the Nature of Reality.* (2010) p.3
8. Gosler, AG: Surprise and the Value of Life, *in Real Scientists Real Faith.* Berry, RJ (Ed) Monarch Books, Oxford (2009) p.184
9. Lennox, JC: *God's Undertaker. Has Science Buried God?* (2007) pp.167–8
10. Floridi, L: *Information: A Very Short Introduction.* (2010) pp.110–111
11. Lloyd, S: *Programming the Universe.* Knopf, NY (2006) p.40 *in* Gregersen, NH: God, matter, and information: towards a Stoicizing Logos Christianity *in* Davies, P & Gregersen, NH (Eds): (2010) p.334
12. Missler, C: *Cosmic Codes.* Koinonia House (2004) p.369
13. A dictum attributed to S. Alexander by J. Kim (Kim, 1992, pp.134–5) *in* Peacocke, A: The sciences of complexity: a new theological resource. Davies, P & Gregersen, NH (Eds): *Information and the Nature of Reality.* (2010) p.254
14. Davies, P: Life force. *New Scientist* 163(2204):27–30 (1999)
15. Davies, P: *The Fifth Miracle.* Simon and Schuster, NY p.259 (1999)
16. Das Wesen der Materie (The Nature of Matter), 1944 speech in Florence, Italy. Source: Archiv zur Geschichte der Max-Planck-Gesellschaft, Abt. Va, Rep. 11 Planck, Nr. 1797 *as quoted in:* Braden, Gregg *The Spontaneous Healing of Belief* Hay House (2008) p.212 [en.wikiquote.org/wiki/Max_Planck – accessed 15 Oct 2016]
17. Wiles, RB: *The Information Paradox.* Information Press (2014) pp.236–7

18. Bradley, WL: Nature. Designed or Designoid *in* Dembski, WA (Ed): *Mere Creation. Science, Faith and Intelligent Design.* InterVarsity Press (1998) pp.48–9
19. *Ibid* p.40
20. Galileo Galilei: *Il Saggiatore* [The Assayer] 1623 *in* Fowler, TB & Kuebler, D: (2007) p.127
21. Rovelli, C: *Reality is not What it Seems. The Journey to Quantum Gravity.* Allen Lane (2014) p.102
22. Wertheim, M: Are we just numbers in the matrix? *Cosmos* 52 (2013) p.65
23. Behe, MJ: *The Edge of Evolution.* Simon & Schuster, NY (2007) pp.239–240
24. Gonzalez, G & Richards, JW: *The Privileged Planet*. (2004) p.297
25. Davies, P: *God and the New Physics.* Simon and Schuster. NY (1983) p.189
26. Gingerich, O: Dare a Scientist Believe in Design? *In Evidence of Purpose*. Templeton, JM (Ed) Continuum, NY (1994) p.25 *in* Strobel, L: *The Case for a Creator.* (2004) p.128
27. Currivan, J: *The Cosmic Hologram – In-formation at the Center of Creation.* Inner Traditions Rochester (2017) p.193
28. Bohm, D: *Dialogues with Scientists and Stages: the Search for Unity in* Wiles, RB: (2014) p.256
29. Currivan, J: *The Cosmic Hologram – In-formation at the Center of Creation.* (2017) p.106
30. Andrews, E: *Who Made God? Searching for a Theory of Everything.* (2009) pp.105–6
31. Lennox, JC: *Seven Days that divide the world: the beginning according to Genesis and Science*. Zondervan, Grand Rapids (2011) p.89
32. Wald, G: The Origin of Life, *in Life: Origin and Evolution*, WH Freeman, San Francisco (1979) p.24 *in* Sarfati, J: *By Design.* Creation Books (2008) p.190
33. Dawkins, R: *The Greatest Show on Earth.* (2009) pp.416–7
34. Wald, G: The origin of life, *Scientific American*, 191:46 (1954) *in* Sarfati, J: (2008) p.189
35. Hoyle, F: *in* Stockwood, M. (ed), *Religion and the Scientists: Addresses Delivered in the University Church, Cambridge, Lent Term, 1957*, SCM Press: London (1959) p.64
36. Sagan, C: *Pale Blue Dot.* Ballantine Books, NY (1994) p7 *in* Gonzalez, G & Richards, JW: *The Privileged Planet*. (2004) p.247
37. Gonzalez, G & Richards, JW: *The Privileged Planet*. (2004) p.162
38. Trigg, R: *Beyond Matter. Why Science Needs Metaphysics.* Templeton Press (2015) p.141
39. Feser, E: *The Last Superstition. A Refutation of the New Atheism.* St Augustine Press (2008) pp.223–6
40. Lennox, J & McGrath, A: *Science and the God Question.* (DVD) Fixed point Foundation, USA (2007)
41. Davies, P: Interview, *in* Lovett, RA: Looking to the Skies. *Cosmos* 20:81 (2008)
42. US National Academy of Sciences *in* Andrews, E: *Who Made God?* (2009) p.65
43. Singer, P & Lennox, J: *Is there a God* (debate) Melbourne Town Hall, Australia (20 Jul 2011)
44. Bohm, D: *Wholeness and the Implicate Order*, Routledge & Kegan Paul, London, Boston (1980) p.48 http://www.spaceandmotion.com/Physics-David-Bohm-Holographic-Universe.htm
45. Loeb, A: Good data are not enough. *Nature* 539, pp.23–25 (3 Nov 2016)
46. Einstein, A: *Letter to Phyllis Wright* 24 Jan 1936 *in* Jammer, M: *Einstein and Religion: Physics and Theology,* Princeton Univ. Press, NJ (1999). http://quoteinvestigator.com/2011/12/16/spirit-manifest/
47. Hawking, SW: *A Brief History of Time: From Big Bang to Black Holes.* Random House (1988) p.127
48. McGrath, A & JC: *The Dawkins Delusion? Atheist fundamentalism and the denial of the divine.* Society for Promoting Christian Knowledge, London (2007) p.10
49. Lennox, JC: *Seven Days that divide the world.* (2011) pp.94–5
50. Floridi, L: *Information: A Very Short Introduction*. (2010) pp.69–70
51. John 1:1–5; John 7:38; Psalm 36:9; Zechariah 13:1; John 4:10–14 (NIV)

Index

About the Author

Robert Wiles is a senior lecturer in the Rural Clinical School at the Australian National University and is the rural academic co-ordinator for the Snowy Mountains region to the south of Canberra. After graduating from the Faculty of Medicine, University of Sydney, he spent five years in the United Kingdom pursuing post-graduate studies in obstetrics, anaesthetics, orthopaedics and general practice before returning to Australia. He has been a rural doctor in Cooma for over twenty-five years, most of which he has also been involved with undergraduate and post-graduate teaching in rural medicine.

He has been married for thirty years with four adult children and has been involved with Christian schooling for over twenty years.

He has a passion for teaching and learning in Medicine which has extended to exploring and understanding the information that defines our Universe. His previous book, *The Information Paradox*, was published in 2014.

www.ingramcontent.com/pod-product-compliance
Lightning Source LLC
LaVergne TN
LVHW050915210926
848482LV00012B/444